NEWNES POWER ENGINEERING SERIES

Electronic Control of Switched Reluctance Machines

NEWNES POWER ENGINEERING SERIES

Series editors
Professor TJE Miller, University of Glasgow, UK
Associate Professor Duane Hanselman, University of Maine, USA
Professor Thomas M Jahns, University of Wisconsin-Madison, USA
Professor Jim McDonald, University of Strathclyde, UK

Newnes Power Engineering Series is a new series of advanced reference texts covering the core areas of modern electrical power engineering, encompassing transmission and distribution, machines and drives, power electronics, and related areas of electricity generation, distribution and utilization. The series is designed for a wide audience of engineers, academics, and postgraduate students, and its focus is international, which is reflected in the editorial team. The titles in the series offer concise but rigorous coverage of essential topics within power engineering, with a special focus on areas undergoing rapid development.

The series complements the long-established range of Newnes titles in power engineering, which includes the *Electrical Engineer's Reference Book*, first published by Newnes in 1945, and the classic *J&P Transformer Book*, as well as a wide selection of recent titles for professionals, students and engineers at all levels.

Further information on the **Newnes Power Engineering Series** in available from bhmarketing@repp.co.uk
www.newnespress.com

Please send book proposals to Matthew Deans, Newnes Publisher
matthew.deans@repp.co.uk

Other titles in the Newnes Power Engineering Series
Acha, Agelidis, Anaya & Miller   Electronic Control in Electrical Systems   0-7506-5126-1
Agrawal   Industrial Power Engineering and Applications Handbook   0-7506-7351-6

NEWNES POWER ENGINEERING SERIES

Electronic Control of Switched Reluctance Machines

Edited by T J E Miller

Newnes

OXFORD • AUCKLAND • BOSTON • JOHANNESBURG • MELBOURNE • NEW DELHI

Newnes
An imprint of Butterworth-Heinemann
Linacre House, Jordan Hill, Oxford OX2 8DP
225 Wildwood Avenue, Woburn, MA 01801-2041
A division of Reed Educational and Professional Publishing Ltd

A member of the Reed Elsevier plc group

First published 2001

British Library Cataloguing in Publication Data
A catalogue record for this book is available from the British Library

ISBN 0 7506 50737

Typeset in 10/12pt Times Roman by Laser Words, Madras, India
Printed and bound in Great Britain by MPG Books Ltd, Bodmin, Cornwall

Contents

Preface

The switched reluctance machine could be said to have a pivotal role in the changes in electric machine technology which have been taking place since the mid-1960s, when digital and power electronics and computerized design methods suddenly started to expand. Although the commercial production of switched reluctance machines is still very small, it is easy to appreciate why it generates so much attention. It sits squarely at the centre of the quest for a new balance between the copper and iron of the machine, and the silicon of the drive. As new boundaries are unfolding in power electronic devices and digital control, the old world of the electric machine has reacted with new inventions and adaptations, generally along classical lines. But the switched reluctance machine introduces a new balance, in which the copper and iron are diminished in quantity, complexity, and cost, in favour of a greater reliance on sophistication in the controller. Inevitably such a fundamental shift produces side effects (such as the acoustic noise problem) and, in the face of great technical and commercial competition from more conventional technologies, the switched reluctance drive has had limited commercial success. But in every example of successful application of the technology, mastery of the control has been the key factor: the motors look very ordinary – indeed that is the point about them, straightforward to manufacture and low in cost. The logic of this is that as electronics progresses still further, the opportunities for successful application of switched reluctance motors and generators will grow, and although that is also true of conventional technologies, none of them approaches the ratio of added value between the electronics and the machine which is found in the switched reluctance system. Small wonder, then, that the patenting of new inventions in this field is expanding at an extraordinary rate. With this background the need for a basic introductory text on the control of switched reluctance machines is self-evident.

The book is intended for engineers in industry and in the large research community in electric machines and drives. For anyone engaged in the development of reluctance-motor drive systems it is hoped that it will serve as a useful reference work. Since it is written from first principles, it can be used for studying the subject from scratch, although there are no problems and only a few specific worked examples, so it is not intended as a course text (the subject is in any case too specialized for university courses). A basic understanding of the switched reluctance motor and its control can be obtained by reading only Chapters 3 and 5, which present the basic electromagnetics, performance characteristics, and control requirements. Chapter 4 is recommended as an introduction to the noise question, and Chapters 6 and 8 as examples of the maximum levels of sophistication normally brought to bear in the controller. For the history up to the present time, including comments on the status of patents, read Chapter 2; for sensorless control, Chapter 7; for test methods and electronic implementation of the controller, Chapter 9; and for generator operation, Chapter 10.

The bibliography contains over 200 references. Reflecting the contents of the book, these cover much of the early history together with a selection of more recent articles. However, it is by no means comprehensive.

T.J.E. Miller

Acknowledgements

This book arises largely from the work of the *SPEED Laboratory* at the University of Glasgow in the period from about 1987 to 2000. Several of the authors are or were employed in the *SPEED Laboratory* or on the University staff, or as research engineers or graduate students, and the others have all worked closely with the *SPEED Laboratory* and some of its member companies. We would like to thank all the engineers in *SPEED* companies who have helped us with our work over several years, and also to acknowledge others in the field who made pioneering advances and laid foundations which made this work possible: especially Professor Peter Lawrenson and Dr Michael Stephenson at SR Drives Ltd, Professor John Byrne of University College, Dublin, and Professor Martyn Harris of the University of Southampton. More recently these interactions have included Professor Y. Hayashi of Aoyama-Gakuin University, Tokyo; Mr Shinichiro Iwasaki of Aisin Seiki Company, Kariya, Japan; J.R. Hendershot Jr at Motorsoft, Professor Mehrdad Ehsani of Texas A&M University, Professor Ion Boldea of the Technical University, Timisoara Romania, and several colleagues at the General Electric Company, Emerson Electric, ITT Automotive (now Valeo), National Semiconductor, Lucas Aerospace, TRW, AlliedSignal, DaimlerBenz, Danfoss, Delphi, Eaton Corporation, Emotron, Ford, Honeywell, Kollmorgen, NSK, Oriental Motor, Picanol, Hamilton Sundstrand, Tecumseh Products Company, Dana, A.O. Smith and Tridelta. Special thanks are due to Peter Miller, Ian Young, Jimmy Kelly and Saffron Alsford of the *SPEED Laboratory* and the university, for their long and exceptional service to our experimental and contract programme; and to Malcolm McGilp, the architect of the *SPEED* software. We would also like to thank Siân Jones and Marjorie Durham of Arnold Publishers. We would also like to acknowledge NSK Company, Japan, for support of Dr Sawata during his PhD at Glasgow, particularly Mr Y. Yamaguchi and Mr T. Inomata; also Consejo Nacional de Ciencia y Tecnologia Mexico (CONACyT) for support of Dr Gallegos-Lopez; and the University of Aalborg and subsequently the Engineering and Physical Sciences Research Council for support of Dr Kjaer. We would also like to mention Alan Hutton, Peter Bower, Ken Evans, Roger Becerra, and Duco Pulle who helped us in the early days.

For permission to use previously published material we would like to thank the Institution of Electrical Engineers, the Institute of Electrical and Electronics Engineers, the

Smithsonian Institution, McGraw-Hill Book Company (Figure 2.10), English Universities Press (Figure 2.11), and Elektrische Bahnen (Figures 2.18 and 2.19).

Every effort has been made to obtain permission for the re-use of copyright material in this book, but in a few cases we were not able to trace the copyright holder. In these cases we would be pleased to add acknowledgements in future reprints or new editions.

Abbreviations

a.c.	alternating current
ADC	analogue-to-digital converter
A/D	analogue-to-digital
ALA	axially laminated anisotropic
ASIC	application-specific integrated circuit
CGSM	current-gradient sensorless method
CHA	channel A
d.c.	direct current
DAC	digital-to-analogue converter
D/A	digital-to-analogue
DSP	digital signal processor
EPROM	electrical programmable read-only memory
EMF	electromotive force
FPGA	field-programmable gate array
I/O	input/output
MMF	magnetomotive force
PC	personal computer
PI	proportional/integral
PM	permanent magnet
PSD	power spectral density
PWM	pulse-width modulation
r.m.s	root-mean-square
TPU	timer-processor unit
TSF	torque-sharing function
ZVL	zero-volt loop

1

Introduction

T.J.E. Miller
SPEED Laboratory, University of Glasgow

1.1 Definition and properties

1.1.1 General definition

A *reluctance motor* is an electric motor in which torque is produced by the tendency of its moveable part to move to a position where the inductance of the excited winding is maximized. The motion may be rotary or linear, and the rotor may be interior (as in Figure 1.1) or exterior. Generally the moveable part is a simple component made of soft magnetic iron, shaped in such a way as to maximize the variation of inductance with position. The geometrical simplicity is one of the main attractive features: since no windings or permanent magnets are used, the manufacturing cost appears to be lower than for other types of motor, while the reliability and robustness appear to be improved.

1.1.2 Synchronous and switched reluctance motors

The definition above is broad enough to include both the *switched* reluctance motor and the *synchronous* reluctance motor, Figure 1.2. The idealized forms of these machines are defined and compared in Table 1.1. Both machines are the subject of extensive academic and industrial research, but are not very closely related in the literature or the laboratory or the factory.

1.1.3 Relationship with VR stepper motors

The switched reluctance motor is topologically and electromagnetically similar to the variable-reluctance stepper motor (Acarnley, 1982). The differences lie in the engineering design, in the control method, and in performance and application characteristics. The switched reluctance motor is normally operated with shaft-position feedback to synchronize the commutation of the phase currents with precise rotor positions, whereas the stepper motor is normally run *open loop*, i.e. without shaft-position feedback. Whereas switched reluctance motors are normally designed for

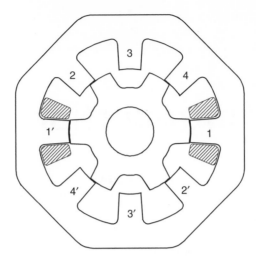

Fig. 1.1 Cross-section of four-phase switched reluctance motor with 8 stator poles and 6 rotor poles. Each phase winding comprises two coils, wound on opposite poles.

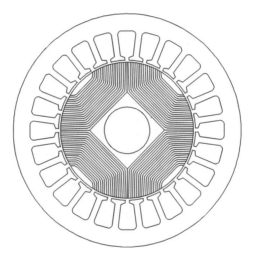

Fig. 1.2 Axially laminated synchronous reluctance motor. This is a true a.c. rotating-field motor. The stator and its polyphase winding are essentially the same as in the induction motor, and the supply is sinusoidal.

efficient conversion of significant amounts of power, stepper motors are more usually designed to maintain *step integrity* in position controls.

1.1.4 Terminology

The first use of the term *switched reluctance motor* appears to have been by S.A. Nasar (1969) to describe a rudimentary switched reluctance motor. Interestingly, Nasar referred to his motor as a d.c. motor, which echoes the 'brushless d.c.' motor. In the United States the term *variable reluctance motor* (VR motor) is often preferred.

Table 1.1 Comparison of switched and synchronous reluctance machines

Switched reluctance	Synchronous reluctance
1. Both stator and rotor have salient poles.	1. The stator has a smooth bore except for slotting.
2. The stator winding comprises a set of coils, each of which is wound on one pole.	2. The stator has a polyphase winding with approximately sine-distributed coils.
3. Excitation is a sequence of current pulses applied to each phase in turn.	3. Excitation is a set of polyphase balanced sinewave currents.
4. As the rotor rotates, the phase *flux-linkage* should have a triangular or sawtooth waveform but not vary with current.	4. The phase self-inductance should vary sinusoidally with rotor position but not vary with current.

However, the 'VR' motor is also a form of stepper motor, so this term does nothing to differentiate the switched reluctance motor from its close relative. Widespread use of the term *switched reluctance* in connection with the modern form of this motor is no doubt partly attributable to Professor Lawrenson and his colleagues at SR Drives Ltd (now part of Emerson Electric), (Lawrenson *et al.*, 1980). The terms *brushless reluctance motor* and *electronically commutated reluctance* motor have also been used occasionally to underline the fact that the motor is brushless (Hendershot, 1989; Hancock and Hendershot, 1990). The term *switched reluctance* does not mean that the reluctance itself is switched, but refers to the switching of phase currents, which is an essential aspect of the operation. This switching is more precisely called *commutation*, so *electronically commutated reluctance motor* is perhaps an even more precise term than *switched reluctance motor*. It also draws a parallel with the electronically commutated permanent-magnet motor (i.e. the square-wave or 'trapezoidal' brushless d.c. motor). In both cases the switching performs a function similar to that of the commutator in a d.c. motor.

1.1.5   General characteristics

The switched reluctance motor has no magnets or windings on the rotor. This is both an advantage and a disadvantage. On the one hand, it means the rotor is of simple construction and the costs and problems associated with permanent magnets (such as magnetization and demagnetization) are avoided. On the other hand, the excitation provided by magnets is very considerable in small motors, and this leaves the switched reluctance machine at a disadvantage. The reliance on a single excitation source, coupled with the effects of fringing fields and magnetic saturation, renders the reluctance motor nonlinear in its control characteristics. To achieve performance competitive with that of other modern motor types, it is generally necessary to control the current waveforms electronically: otherwise the efficiency, the noise level, and the utilization of converter volt-amperes can be disappointing. In the past 30 years or so the development of digital and power electronics has made it possible to exploit the characteristics of reluctance machines sufficiently well so that several successful products are now manufactured, and there continues to be a substantial research effort into the development of new ones.

Before about 1965, the variable-reluctance principle was used for many years in various devices, and the rich inventive history of these machines is reflected in Chapter 2 all the way back to the dawn of electrical engineering, before 1840. When the power transistor and digital integrated circuits became available, new designs were developed with improved performance. The fundamental theory of reluctance motors was extended, and the design process was helped by the development of efficient finite-element and simulation software.

During the development period of the modern switched reluctance machine, since about 1965, other competing types of electric machine and drive system have also made great progress: particularly induction motors with field-oriented control and brushless permanent-magnet motors and drives. Both these types are produced in far greater numbers than switched reluctance motors, and for a wider range of applications. This is unlikely to change significantly, and the reluctance machine will most likely remain a specialty machine chosen for its special advantages. Like the other machines, it shares the modern characteristic of mechanical simplicity combined with electronic sophistication.

1.2   The purpose and structure of this book

The purpose of this book is to present an up-to-date account of some of the main control techniques applied to switched reluctance machines. Too much has happened in the past 10–15 years to be recorded in a single book, but on the other hand the subject of switched reluctance machines has little coverage in textbooks and only one (Miller, 1993) is devoted exclusively to the subject. The book is mainly about control, and contains little on the machine design or the calculation of its performance. It is also focused on 'mainstream' control techniques and does not begin to do justice to the enormous variety of special inventions, although it can be observed that most of the successful applications of switched reluctance machines to date are more or less 'conventional' in their basic configuration.

The review in Chapter 2 spans the entire history from the beginning of the age of electric machinery. Reluctance machines were among the first types to be invented and applied, and their history is interesting as a minor theme in the general development of electrical engineering, and important in laying the foundations of the modern subject. In the mid-1960s reluctance machine technology received a new stimulus with the invention of the power transistor – so much so that the previous era appears by comparison to have been one of almost complete dormancy.

In Chapter 3 the electromagnetic principles are reviewed along classical lines. The oversimplified magnetically linear model is used to provide a structure for understanding the commutation of the phases, and the magnetically nonlinear model is then developed to bring out the energy conversion capabilities and explore the performance characteristics.

Chapter 4 presents an example of a motor designed for low noise, together with some of the fundamental principles used in the reduction of acoustic noise. Its location in the book reflects the basic importance of attention to the noise problem, as in many other branches of engineering, but it also exemplifies a modern invention, the *stagger-tooth motor*[TM], which (among many others) reflects the continuing innovation in the field of reluctance machines in spite of the maturity of the underlying concepts. This chapter also acts as a pivot between the earlier chapters, which are dominated by the characteristics of the machine, and the later chapters, which are almost wholly concerned with the drive and its control algorithms. For most of Chapter 4 it is assumed that the controller is an *average torque* controller.

The account of electronic control starts in earnest in Chapter 5 at the simplest level of average torque control, which is concerned with the control of the current waveform through a small number of parameters such as the turn-on and turn-off angles and the current reference. This leads naturally to Chapter 6 which explores the possibilities for instantaneous torque control, to assess the potential of the switched reluctance motor as a servo motor in comparison with an induction motor with field-oriented control. Here the objectives are to produce smooth torque and rapid dynamic response, while overcoming the natural nonlinearities of the machine from a control point of view.

Chapter 7 is a comprehensive review of sensorless control methods. The development of controls which avoid the use of physical shaft position sensors for commutation has been a strong theme in the development of switched reluctance machines, partly because of the perception that encoders and resolvers are costly, and partly arising from the quest for machines that can survive exceptionally hostile environments. The subject of sensorless control therefore includes 'low-cost' embodiments with the smallest possible part count, and more sophisticated schemes with appreciable computational sophistication.

Chapter 8 describes a particular example of a drive system designed for very low torque ripple and low acoustic noise, with exacting dynamic control requirements in a machine with high torque density, developed for an electrically assisted automotive steering system. Unlike Chapter 4 which is built on an electromagnetic invention, this system is built on electronic invention and uses a more or less conventional switched reluctance motor, which it treats as a 'black box'. The sophistication of the controller is brought out through the mathematical development of the control theory and a system model.

Chapter 9 describes methods for system development and test, at a detailed electronics level, and Chapter 10 specializes the discussion to the switched reluctance machine as a generator. The generating capability of the switched reluctance machine is less widely known than the motoring capability, and there is more design freedom arising from the absence of self-starting requirements. As a 'fault-tolerant' system it has considerable appeal, and this aspect is investigated.

2

Development history

Antony F. Anderson

2.1 Introduction

The development of semiconductor rectifier and power switching technology in the early 1960s led to its rapid and successful application to variable speed drives. Few of these drive systems were new in principle, but certain system elements – such as motor generator sets, mercury arc converters, magnetic amplifiers etc. – could now be replaced by solid-state electronics, that performed the same functions with considerable cost and performance benefits. The existence of solid-state power switching technology also stimulated an interest in possible alternative and simpler motor/control configurations, of which the switched-field reluctance motor, long out of favour, was one.

In the last 30 years a variety of configurations of reluctance motor and switching methods have been investigated, both theoretically and experimentally. Various forms of switched reluctance motor based on the most significant patents are now manufactured under licence. According to one source, 67 patents were published worldwide on switched reluctance motors before 1976 and 1755 between 1976 and the end of 1999.[1] The same source estimates the total number of papers published before 1976 to be 11 and between 1976 and the end of 1999 to be 1847.

The statistics certainly suggest that by far the greatest academic and industrial activity on switched-field reluctance motors has taken place post 1976, but the figures do not in any way do justice to the quality and significance of the work carried out prior to that date. The apparent lack of earlier material suggested by these figures arises for two reasons. Firstly, earlier patents and papers have had plenty of time to get buried and forgotten. Secondly, relevant material tends to be found in a variety of unlikely places. It is worth noting also that the number of patents, whilst it may well give some indication of interest in the subject, will not necessarily bear any relationship to the importance of the work carried out: earlier work, although much less plentiful, is more likely to contain significant elements of innovation.

Priority in the switched reluctance motor field is by no means clear cut, because the switched reluctance motor has a very long, and largely forgotten pre-history, with a multiple inheritance from d.c. and a.c. motor, stepper motor and inductor alternator ancestors, not to mention an inheritance from other electromagnetic devices,

---

[1] Fleadh Electronics Database of Switched Reluctance Motor Patents at: http://www.fleadh.co.uk

such as electric clocks, vibrators and timing mechanisms. Relevant prior art therefore may be found in many different places. It is a criticism that may be justly levelled against some recent patents and some otherwise excellent published papers that they fail to position themselves adequately in the context of this somewhat dispersed prior art.

This chapter will therefore briefly outline salient aspects of the historical development of the switched-field reluctance motor and provide a framework of reference in which to assess the innovative elements of the more recent developments described in subsequent chapters.

2.2 Early examples

2.2.1 The electromagnetic engine: an early reluctance motor

Reluctance motors were among the earliest electric motors to be developed (1830s–1850s).[2] Their origins lie in the horseshoe electromagnet of William Sturgeon (1824)[3] (Figure 2.1) and the improved version of Joseph Henry (Figure 2.2) and in attempts to convert the 'once only' attraction for an iron armature into oscillatory or continuous motion.

The Rev. William Ritchie (1833), Professor of Natural Philosophy in the Royal Institution and in the University of London, knew of Henry's improvement to Sturgeon's electromagnet and of the considerable lifting powers obtained. He realized intuitively that if the electromagnet was to be put to practical use some kind of model for its behaviour would be required. He seems to have been the first to attempt to establish

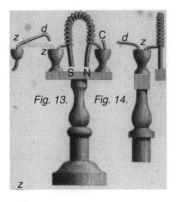

Fig. 2.1 Sturgeon's electromagnet, 1824. *Transactions of the Society of Arts, Manufactures & Commerce,* Vol. XLIII, 1825. Sturgeon was awarded a premium of £25 by the Society in 1824.

---

[2] Called electromagnetic engines at the time.

[3] Sturgeon, W.: 'Improved Electro Magnetic Apparatus', *Trans. Soc. Arts, Manufactures & Commerce,* Vol. XLIII, 1825, pp. 37–52, plates 3 & 4.

Fig. 2.2 Henry's large electromagnet constructed in 1831 for Benjamin Silliman Sr of Yale College. Now in the Smithsonian Institution. This magnet was capable of supporting a weight of 1000 lbs. Henry's electromagnet has a multilayer winding of fine wire and a relatively short magnetic circuit. The power of Henry's magnet to attract an armature of iron fired the imagination of others to build the first crude electric motors. (Courtesy of the Smithsonian Institute).

some kind of magnetic circuit law.[4] Like most of his contemporaries, he was disadvantaged by being unaware of Ohm's discovery of the law of conduction for electric circuits (1827), which might have provided him with a useful analogy.[5,6]

Ritchie wrote: 'Though the astonishing lifting powers of electro-magnets be sufficiently known, yet no attempt seems to have been made to investigate the law, if such exists, which connects this power with the length of the magnetic circuit ...' Although he failed to establish the magnetic circuit law, he made a very interesting observation:

Having constructed, with great pains, an electro-magnet according to the American method, and connected it with a battery, I found it would carry about one hundred and forty pounds. I then rolled about twelve feet of copper ribbon about the middle

[4] Ritchie, W.: 'Experimental Researches in Electro-Magnetism and Magneto-Electricity', *Phil. Trans.*, 1833 [2]: 313–321.

[5] Weber, R.: 'More Random Walks in Science', Institute of Physics 1982: 198, quoting 'Makers of Science', Oxford University Press, 1923, p. 243, says that Ohm's law was described by one of Ohm's German contemporaries as: 'A web of naked fancies' and the German Minister of Education said: '... a physicist who professed such heresies was unworthy to teach science.'

[6] Ohm's law did not become generally known until Jacobi applied it in his paper on the application of electromagnetism to the movement of machines (1835) which became available in English translation in *Taylor's Scientific Memoirs* published in London in 1837.

of the lifter, which weighed about half a pound, and connected the ends of the coil with the same battery, the electromagnet being now used as the lifter, and was surprised to find that the lifter was a more powerful magnet than that which had cost me so much labour to prepare. All that is necessary, then, in making a powerful electro-magnet, is simply to roll a ribbon of copper (metallic contact being prevented by a thin tape interposed) about a short bar of soft iron, and use a short-shoe lifter.

Ritchie's highly significant conclusion that magnetic circuits should be kept short and, by implication, well coupled with the electric circuit, was largely lost on his contemporaries.[7] Early electric motors, of which there was an amazing variety, were largely designed by trial and error and, in the main, had very poor magnetic circuits.[8] It was not until John Hopkinson's work on the magnetic circuit (1886) that electrical machine design could be carried out systematically and only then that the importance of keeping the magnetic circuit short began to be appreciated.[9]

Many early motor designs were reluctance motors and these were strongly influenced by the steam engine, with electromagnets, armatures and current switching arrangements being regarded as the electromechanical equivalents of cylinders, pistons and valve gear. It is neither surprising that many of these early machines looked like steam engines nor that they were called 'electromagnetic engines'.

Particularly interesting early designs, from a topological point of view, were the two small eccentric electromagnetic engines built for Charles Wheatstone by the self-taught instrument-maker and pioneer submarine cable manufacturer William Henley[10] circa 1842,[11] which incorporated the ideas contained in Wheatstone's patent 0922 of 1841 (Figures 2.3, 2.4).

Generally these early motors may be classified as: motors with separate sequentially switched magnetic circuits (variable self-inductance); motors with mutually coupled sequentially switched magnetic circuits (variable mutual inductance). The former type were in the majority, but a significant minority of designs fall into the latter category.

- Examples of the former type in which the magnetic circuits are not coupled were the electromagnetic engines of Callan,[12] Davidson[13] (Figure 2.5) and Taylor[14] (Figure 2.6) in which two or more stationary electromagnets were energized

---

[7] Ritchie died of a fever caught in Scotland in 1837 (DNB). Had he lived, he would undoubtedly have made a very significant contribution to the development of electrical machines.

[8] For example, Edison's dynamos of the 1880s with their long thin field electromagnets show the end of a line of a particular line of evolution away from good magnetic circuit design.

[9] Hopkinson, J.: 'Dynamo Electric Machinery'. *Phil. Trans. Royal Soc.*, Part 1, 1886, also Hopkinson, J.: *Collected Papers*, Cambridge University Press, 190, Vol. 1, pp. 84–121. In this work Hopkinson for the first time lays out the Magnetic Circuit Law and applies it to the design of dynamo-electric machinery.

[10] Anderson, A.F.: 'William Henley, pioneer instrument maker and cable manufacturer 1813 to 1882'. *Proc. IEE*, Vol. 132, Pt. A, No. 4, July 1985, pp. 249–261.

[11] Bowers, B.: *Sir Charles Wheatstone*, Science Museum, 1975, pp. 72–85.

[12] McLaughlin, P.J.: 'Nicholas Callan Priest Scientist 1799–1864', published by Clonmore & Reynolds and Burns Oates 1965. Rev Nicholas Callan of Maynooth built a number of reluctance motors in the late 1830s which are still to be seen in his laboratory at Maynooth College, Eire.

[13] 'Davidson's Electromagnetic Engine': *The Penny Mechanic and Chemist*, 23 September 1843, pp. 298–299 and 30 September 1843, p. 305.

[14] 'Taylor's Electro-Magnetic Engine': *Mechanic's Magazine*, No. 874, Saturday 9 May 1840.

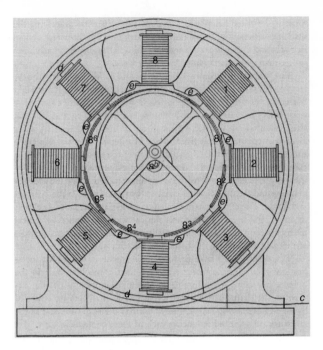

Fig. 2.3 Wheatstone eccentric electromagnetic engine – type 1, UK patent 9022 of 1841. Built by W.T. Henley.

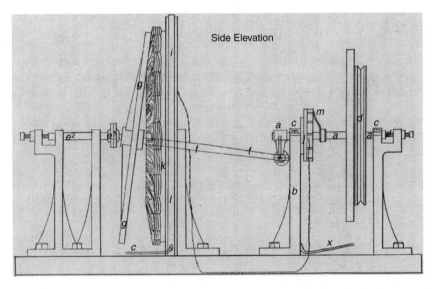

Fig. 2.4 Wheatstone eccentric electromagnetic engine – type 2. Patent 9022 of 1841. Built by W.T. Henley.

sequentially to produce tangential forces on a series of armatures laid on the surface of a rotating drum. All three inventors were building switched-field reluctance motors entirely independently in the period 1837–40. Two problems arose with this type of motor: the first was that at the end of the switching cycle all the stored

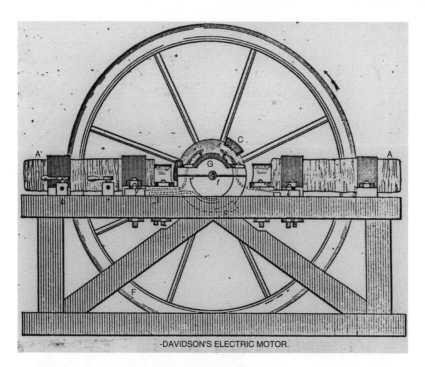

Fig. 2.5 Davidson's motor. Built circa 1839. From the *Penny Mechanic and Chemist*, 30 September 1843: 305. Demonstrated Aberdeen 1840, Edinburgh 1841 and London 1842. Showing two electromagnets A and A' and one of three axial armature bars mounted on the rotor. Drawing by W.T. Henley.

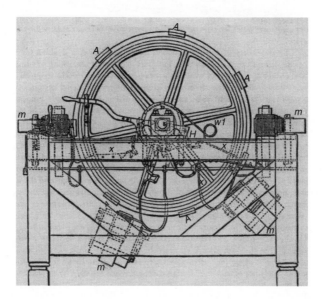

Fig. 2.6 Taylor's motor – from the *Mechanics Magazine*, 9 May 1840. Similar in principle to Davidson's motor. Drawn here with seven equally spaced armature bars and four electromagnets. There is provision for advancing or retarding the coil switching. (Note: the apparently arbitrary arrangement of the two lower electromagnets suggests that they may have been misdrawn to deter imitators.)

inductive energy in the electromagnet had to be dissipated, with resultant arcing and sparking – often spectacular – at the switch[15]; the second was the difficulty of making the stator frame sufficiently strong to withstand the radial out-of-balance magnetic forces.

- Examples of the latter type of motor are the electromagnetic engines of Charles Page (USA) (Figure 2.7). Page eventually built a 16 HP electromagnetic engine (1851) for a battery driven electric locomotive that achieved a speed of 19 miles an hour on the Baltimore and Ohio Railway.[16] Page's motors used opposing sets of solenoids arranged in tandem rows that were successively energized to pull dual axial iron bars, first one way and then the other.[17] The main advantage of the solenoidal construction was that the electromagnetic forces were in the direction of motion and there were no out-of-balance forces. The direct successors of such solenoidal coil machines are various linear reluctance actuators and, in particular, the linear variable reluctance-type balanced actuators used for some control purposes today (Figure 2.8).[18]

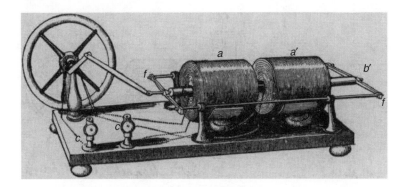

Fig. 2.7 Page's motor – early form (1844) with double-acting solenoids (*American Journal of Science*, 49, 1845, page 133). In Page's later motors, a set of coils was placed close together on a frame to form a long cylinder of sections (coils). The coils were all connected in series, with each coil joint brought out to a segment of a linear 'circuit changer', effectively a linear commutator. Several adjacent coils were energized at the same time with current transferring into the leading coil (rising current) and current transferring from the trailing coil (falling current) via the 'circuit changer'. The iron armature was therefore in a more or less constant travelling field. It is clear that in this type of switched-field reluctance motor both mutual and self-inductances change with armature position.

---

[15] This problem does not arise in the modern electronically switched motor because ways have been devised to feed the stored energy back into the supply.

[16] Post, Robert C.: *Physics, Patents and Politics – A biography of Charles Grafton Page*, Science History Publications, New York, 1976.

[17] In Page's later motors the coils were all connected in series, with each joint connected to a segment of the cut off slide. Three coils were energized at once, with current switch-on occurring in the leading coil and current cut-off occurring in the trailing coil. The armature therefore was in a travelling field.

[18] Some of these have two actuating coils. There are three components of reluctance force. Two relate to rate of change of self-inductance of the individual coils and the third relates to the product of the two coil currents multiplied by the rate of change of mutual inductance between the coils:.

$$\text{Force on armature} = \tfrac{1}{2}I_1^2\, dL_1/dx + \tfrac{1}{2}I_2^2\, dL_2/dx + \tfrac{1}{2}I_1 I_2\, dM_{12}/dx$$

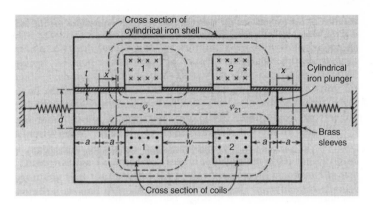

Fig. 2.8 Two-coil solenoid magnet (Figure 2.10, Example 2.6, Fitzgerald & Kingsley, *Electrical Machinery*, 2nd Edition, 1952). The force on the plunger has three components, two related to the rate of change of self-inductance of the two coils with displacement and the third related to rate of change of mutual inductance. *Ibid.* pp. 64–66: full linear analysis of force on plunger against displacement for two-coil mutually coupled system. (Courtesy of McGraw Hill Book Company, Inc.).

2.2.2   The reluctance motor eclipsed by d.c. and a.c. machines

The development of the commutator motor (1870s), the discovery of the magnetic circuit law (in the 1880s), and the realization that both a.c. and d.c. armature windings could be placed in slots, resulted in motors that were manifestly superior in performance to the switched reluctance motor as it then existed. Consequently, well before the end of the 19th century, the switched reluctance motor had ceased to be considered as more than a philosophical instrument-maker's toy. It lived on in a number of special purpose electromagnetic devices which had evolved during the same period, for particular uses where efficiency was not a prime requirement: electromagnetic relays, for amplification of electrical signals; timing and escapement mechanisms, including electric clocks and some instrumentation mechanisms; vibratory devices (electric bells, small paint sprayers and some types of electric razor), rocking mechanisms sometimes used in animated shop displays; reluctance type actuators (valve actuators, contactors etc.) where the requirement was for large forces to act over short distances.[19]

2.2.3   The phonic wheel

The phonic wheel[20] attributed to La Cour (1878)[21] and Rayleigh (1879)[22] is essentially a switched-field reluctance motor. Its invention arose out of the requirement to produce

---

[19] Some of these devices use mechanical commutation.

[20] The phonic motor consisted of a toothed wheel, the teeth of which were attracted in sequence by an electromagnet. The current was supplied to the electromagnet by a make and break switch connected to one of the limbs of an electrically maintained tuning fork: the wheel speed was therefore always dependent on the vibration frequency of the tuning fork and was inversely proportional to the number of teeth on the wheel.

[21] La Cour, M.P.: 'Roue phonique pour la régularisation du synchronism des mouvements,' *Comptes Rendus de l'Académie des Sciences*, 1878, 87, p. 499.

[22] Lord Rayleigh: Societies and Academies (Physical Society) section of *Nature*, 1878, 18, p. 111; *Scientific Papers*, Cambridge University Press, 1899, 1 p. 355; 2 p. 179.

constant speeds of rotation for instrumentation and measurement purposes. The phonic wheel could equally be operated from a d.c. supply with a make-and-break switch driven by a tuning fork, or in synchronous mode from an a.c. supply. An example of a single-phase phonic wheel is shown in Figure 2.9. Two- and even three-phase versions of phonic wheels, with and without polarizing d.c. windings, have been built.[23] The phonic motor is related to the slow-speed synchronous reluctance motor used to drive some electric clocks, the vernier reluctance motor, and the high-frequency inductor alternator.

Walker (1946)[24] in his review of high-frequency inductor alternator types (up to 50 kHz), which spans a period of over 40 years, shows that by that date almost every conceivable combination of double and single saliency and of concentrated and distributed excitation and output windings had been used, in order to obtain high frequencies without the penalty of impossibly small pole pitches. Figure 2.10 shows an example of one such inductor alternator winding arrangement.

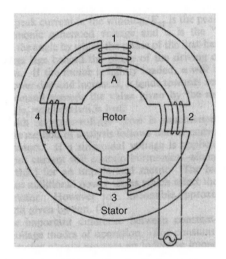

Fig. 2.9 Caro, D.E.: 'The Theory of the Operation of a Phonic Motor', published as *IEE Monograph No. 18*, IEE Measurements Section, 15 December 1951. (Courtesy of Institution of Electrical Engineers).

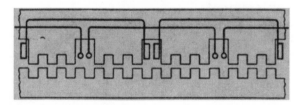

Fig. 2.10 Inductor alternator. From Walker, J.H. 'High Frequency Alternators', *Journal IEE*, Vol. 93, Part II, 1946. (Courtesy of Institution of Electrical Engineers).

---

[23] Caro, D.E.: 'The Theory of the Operation of a Phonic Motor', published as *IEE Monograph No. 18*, IEE Measurements Section, 15 December 1951.

[24] Walker, J.H.: 'Inductor Alternators', *IEE Journal*, Vol. 93, Part II, 1946, pp. 67–80.

2.2.4   Boucherôt's synchronous oscillating reluctance motors

In the early years of the 20th century Boucherôt (France) realized that considerable power could be extracted from mechanically tuned electromagnets energized with alternating current. He built a 40 HP oscillating spring-and-ratchet motor which, to all intents and purposes, was a reborn and fairly compact synchronous electromagnetic engine (Figure 2.11).[25] His ideas were not taken further, largely because these 'oscilla' motors were very noisy: 'like the sound of horses' hooves on the pavé.' Small motors working on the same principle are occasionally found, for example, in electric razors and small paint spray compressors.[26]

2.2.5   The synchronous reluctance motor

The synchronous reluctance motor evolved quite differently from the polyphase induction motor by removing rotor segments to create salient poles. The motor runs up to slip speed as an induction motor, pulls into step with the rotating field by means of the reluctance torque and then operates as a synchronous reluctance motor. This simple form of synchronous reluctance motor enjoys limited application up to medium horsepowers, where cheap and robust constant speed drives are required, or where a number of motors have to be driven at exactly the same speed, e.g. for the synchrolifts used for lifting ships out of the water.

Various attempts have been made over the years to increase the effective degree of asymmetry within the rotor of synchronous reluctance motors in order to improve

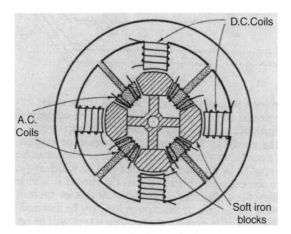

Fig. 2.11 Boucherôt's 40 HP oscillating ratchet motor, showing a.c. and d.c. coils. From Laithwaite E.R., *Propulsion Without Wheels*, English Universities Press, 1966 *Bull. Soc. Int. des Electriciens*, 1908, **18**: 731–735. (Courtesy of EUP).

---

[25] Boucherôt: 'Appareils et machines à courant et mouvements alternatifs', *Bull. Soc. Inst. Des Electriciens*, 1908, 18, pp. 731–735.

[26] It is a moot point whether oscilla motors are synchronous reluctance motors or switched-field reluctance motors fed from a sinusoidal power supply.

specific power output, efficiency and power factor.[27,28] The form of construction must be such as to enable a large flux to be easily established along the main flux path and, at the same time, restrict the possible flux that can be established in directions orthogonal to the main flux path. The most common way of achieving this is by internally slotting the rotor laminations in such a way as to guide the flux along the desired paths.

Kostko first proposed the adoption of a cylindrical rotor with multiple slits along the lines of the direct axis flux in 1923 and this idea is substantially the basis for all current commercial designs of rotor with reluctance slots, flux guides and segments (Figure 2.12). Improvements have been achieved by paying attention to the external rotor shape (use of split-pole or segmental rotors, for example) or by increasing the degree of internal magnetic anisotropy (use of flux guides). It is claimed that some recent designs can more or less equal the induction motor in terms of specific power output and efficiency.

Internal flux guides tend to weaken the rotor structure and limit the degree of magnetic subdivision within the rotor that can be achieved. The axially laminated reluctance rotor (ALA) which uses a strip wound segmental core similar to that used in strip wound transformers allow high ratios of direct to quadrature axis reactance achieves a solid structure whilst still maintaining a high degree of magnetic anisotropy (Figure 2.13).

A high reactance ratio maximizes pull-out torque, but it also tends to make for an unstable motor that does not readily pull into step. This problem may be overcome by feeding the motor from a three-phase variable-frequency supply, with closed-loop

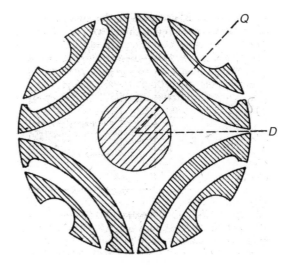

Fig. 2.12 Kostko's reluctance motor rotor with flux guides, 1923. Kostko, 'Polyphase Reaction Synchronous Motors', *J. Amer. Inst. Elect. Engrs*, 1923, p. 1162. (© IEEE, 1923).

[27] Cruickshank, A.J.O. and Anderson, A.F.: 'Development of the Reluctance Motor', *Electronics and Power*, 1966 Vol. 12, p. 48.

[28] Anderson, A.F.: 'A New Type of Reluctance Motor', *IEE Students' Quarterly Journal*, September 1968, pp. 19–24, 45.

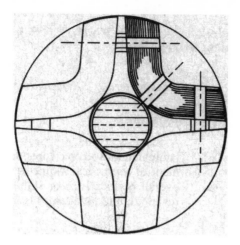

Fig. 2.13 Axially Laminated Rotor 'Cruickshank *et al.*, 1976'. (Courtesy of Institution of Electrical Engineers).

feedback of rotor position. Depending on the type of control used, different motor characteristics can be obtained.[29]

In the synchronous reluctance motor, the rotor normally reacts with the fundamental of the rotating m.m.f. wave which is produced by all the stator winding phases working together. The following points are worth noting:

- Saliency and internal anisotropy are usually confined to one element, the rotor.
- Individual phase windings are distributed, often with nearly, or fully pitched coils.
- The rotating m.m.f. wave is composed of the m.m.f. contributions of individual stator phase windings working together.
- All three phase windings are mutually coupled and interact with the rotor and contribute to the torque.

If the rotor has a degree of internal anisotropy, then by using a pole-changing winding the rotor should be able to operate at integral submultiples of the synchronous speed. Figure 2.14[30,31] illustrates the case of an axially laminated rotor in a stator field having twice the number of poles to the rotor and therefore running at half speed.

An interesting example in which this effect is exploited is the Reluctance Frequency Changer of Forbes (1932)[32] of which a three-to-one 60/180 Hz version is shown (Figure 2.15). This has a two-pole excitation winding (60 Hz), a 2-pole segmented rotor and a 6-pole output winding (180 Hz). These Reluctance Frequency Changers were used to provide high-frequency three-phase supplies for driving woodworking machinery.

---

[29] It is entirely reasonable to categorize a synchronous reluctance motor when it is fed from a pulsed electronic supply as a member of the switched-field reluctance motor family.

[30] Cruickshank, A.J.O., Menzies, R.W. and Anderson, A.F.: *Proc. IEE*, Vol. 113, No. 12, December 1966, pp. 2058–2060.

[31] Cruickshank, A.J.O. and Anderson, A.F.: UK Patent 1, 114,562, 1968.

[32] Forbes, A.W.: 'Simplified Frequency Changer', *Electrical World*, 23 January 1932, pp. 192–194.

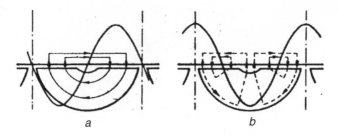

Fig. 2.14 Axially laminated rotor in stator field having twice the number of poles: (a) rotor in direct-axis position; (b) rotor in quadrature-axis position (Cruickshank *et al.*, 1966). (Courtesy of Institution of Electrical Engineers).

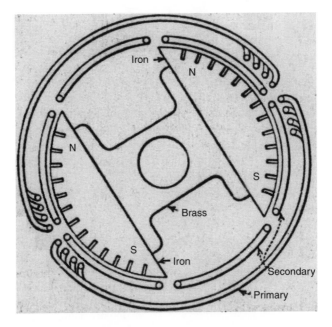

Fig. 2.15 1:3 Reluctance frequency changer. Forbes, 'Simplified Frequency Changers', *Electrical World*, 23 January 1932, p. 192. (Courtesy of Electrical World).

2.2.6   The Admiralty 'M' type stepper motor

The Admiralty 'M' type stepper motor (1924)[33] (Figure 2.16) represents yet another stage of the reluctance motor story. It was developed by Clausen just after the First World War, as a stop-gap measure until the Magslip came along, as a remote servo indicator for naval guns. It overcame the problems of the five different types of step-by-step motors used by the British Navy during the 1914–18 war. Typically, the stepping motors it replaced were of the doubly salient 6-pole/2-pole type, giving 12 steps

---

[33] Clausen, H.: UK Patent 292663, 1924, 'Improvements in step by step motors'.

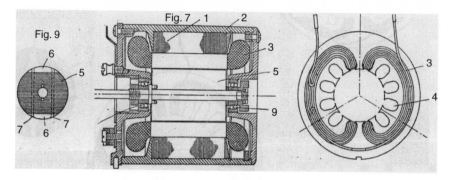

Fig. 2.16 Clausen's 'M' type motor – two pole version. Improvements in step by step motors, British Patent 292663, 1924.

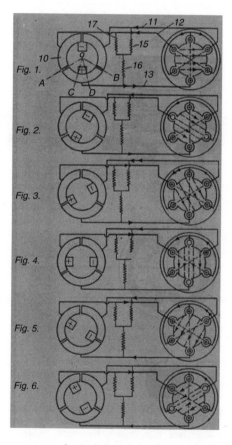

Fig. 2.17 Remote switching arrangement for the 'M' type motor. Improvements in step by step motors, British Patent 292663, 1924. Three-phases are star-connected and are energized either two windings in series, or two windings in parallel connected with the third winding in reverse series via the three part commutator to give a stepped rotating field.

per revolution.[34] The M type motor itself comprised a three-phase, fully pitched stator winding, of a kind typical for small synchronous or induction motors, and an axially laminated reluctance rotor of interleaved iron and aluminium, with a high ratio of direct to quadrature axis reluctance. The stator windings were connected – one phase in series with the other two reversed – by means of a remote switching arrangement, shown in Figure 2.17. The torque resulted from the combined excitation currents in all three phase windings.

According to Clausen, the numbers of 'M' type motors manufactured ran into millions and the motor was still in use in the 1950s during the Korean War. The significance of the fact that it was designed as a replacement for existing concentrated coil, salient-pole stepper motors and used standard synchronous a.c. stator winding technology should be noted.

The 'M' motor design shows quite clearly that by 1924 there had already been a convergence of stepping motor and synchronous motor technology. In essence the patent is seminal (a) for synchronous reluctance motors, because it breaks away from the convention of salient-pole rotors and uses an axially laminated, anisotropic rotor, and (b) because the motor can be seen as the precursor of all subsequent switched reluctance motors with fully pitched and mutually coupled stator windings.

2.3   Early electronic switching in the 1930s using mercury arc rectifiers and thyratrons

The possibilities of using electronic switches to replace the commutator in electrical machines were first demonstrated in the 1930s with a highly sophisticated mercury arc cycloconverter locomotive built by Brown Boveri for the Swiss railways.[35] (Figures 2.18 and 2.19). A six-anode mercury arc rectifier with control grids was used instead of the more usual commutator. In the United States a different approach using thyratrons was used to control large motors.[36,37,38] Here a 400 HP commutatorless fan motor was installed at the Logan power station of the Appalachian Electric Power Company. This motor ran directly from a three-phase supply fed through 18 thyratrons. Reliability proved a problem: there were 20 thyratron failures in 8272 hours of operation.

These two pioneering applications of electronic commutation were both intended to overcome the commutation problems inherent with large variable-speed commutator machines. Lack of reliability and the high cost of the electronic switching devices resulted in both experimental projects being terminated.

---

[34] Clausen, H.: 'Notes on Step by Step Transmission Systems', Evershed and Vignoles' monograph, 1962. The stator windings of the earlier designs were concentrated, with two poles connected in series to give three windings displaced by 120 degrees. Either one or two windings were energized at any given time, depending on the transmitter switch position.

[35] Kern, E.: 'Der kommutatorlose Einphasen-Lokomotivmotor für 40 bis 60 Hertz', *Elektrische Bahnen*, 1931, Vol. 7, pp. 313–321.

[36] Willis, C.H.: 'The Thyratron Commutator Motor', *General Electric Review*, Feb. 1993, p. 76.

[37] Alexanderson, E.F.W. and Mittag, A.H.: 'Thyratron Motors', *Electrical Engineering*, No. 53, Nov. 1934, p. 1517.

[38] Beiler, A.H.: 'The Thyratron Motor at the Logan Plant', *AIEE Trans.*, Vol. 57, Jan. 1938, pp. 19–24.

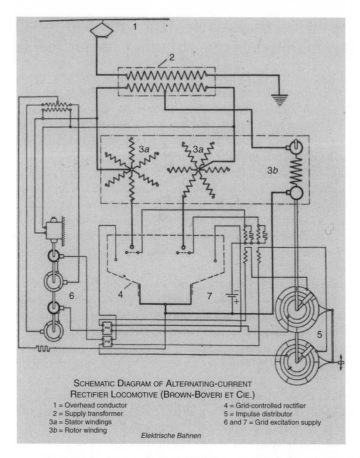

SCHEMATIC DIAGRAM OF ALTERNATING-CURRENT
RECTIFIER LOCOMOTIVE (BROWN-BOVERI ET CIE.)

1 = Overhead conductor   4 = Grid-controlled rectifier
2 = Supply transformer   5 = Impulse distributor
3a = Stator windings   6 and 7 = Grid excitation supply
3b = Rotor winding

Elektrische Bahnen

Fig. 2.18 Schematic of AC mercury-arc commutatorless traction motor (Brown Boveri et Cie, 1931). *Elektrische Bahnen*, 1931, Vol. 7, p. 313. (Courtesy of Elektrische Bahnen).

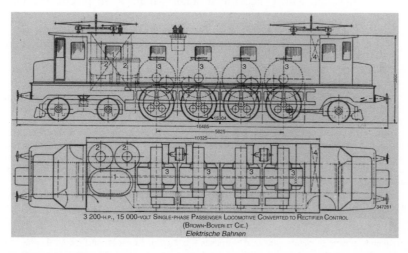

3 200-H.P., 15 000-VOLT SINGLE-PHASE PASSENGER LOCOMOTIVE CONVERTED TO RECTIFIER CONTROL
(BROWN-BOVERI ET CIE.)
Elektrische Bahnen

Fig. 2.19 3 200 HP, 50 000 volt single-phase passenger locomotive converted to rectifier control (Brown Boveri et Cie). *Elektrische Bahnen*, 1931, Vol. 7, p. 313. (Courtesy of Elektrische Bahnen).

2.4 The arrival of the thyristor in the early 1960s: development of the swinging-field machine

The arrival of the first thyristors and power transistors in the early 1960s led to a re-examination of a wide variety of possible d.c. and variable-speed drive configurations. Nevertheless, the switching arrangements in these drives – many being three-phase inverter drives – tended to be rather complex. It was the likely cost of this complexity, in terms of control and switching device configurations, that led Dr Arthur Cruickshank[39] to start an investigation (1961–65) into the possibilities of combining electronic switching using thyristors with the reluctance motor: a motor that is cheap to produce and which requires fewer switching circuits than other types of machine. He realized that two doubly salient reluctance motors could be combined to create a 'swinging-field' machine, in which two sets of stator poles in quadrature could be energized alternately by two switched power supplies using naturally commutated thyristors fed from a three-phase supply. This type of switching circuit is essentially a naturally commutated cycloconverter.

An elemental swinging-field machine with two sets of 4-pole stator winding using thyristor commutation was built (1961–62) with the thyristors fired initially by a somewhat crude mechanical switch mounted on the motor shaft.[40] Electronic control of thyristor switching was introduced and two-path swinging-field and full rotating field switched reluctance drives, using both open- and closed-loop control, were tested in the period 1962–65.[41,42] The basic principles of the switching circuits used and the voltage and current waveforms for two-path and four-path switching are shown in UK Patent 1,114,561 (1968)[43] see Figures 2.20, 2.21, 2.22 and 2.23.

A variety of different stator winding distributions were evaluated theoretically in combination with different rotor designs[44] leading to the conclusion that stator m.m.f. distribution needed to be matched to the type of rotor used. For example, fully pitched concentrated-turn stator windings combined with a near fully pitched segmental or anisotropic rotor might, in theory, give a better switched-field performance than a doubly salient machine with concentrated stator windings. In practical terms, rotor

---

[39] Reader in Electrical Engineering, Queens College Dundee, then part of St. Andrews University.

[40] Anderson, A.F.: Honours Year Thesis 1962, Department of Electrical Engineering, Queens College Dundee.

[41] Anderson, A.F.: 'The Thyristor Control of a Reluctance Motor: Reluctance Motor of Improved Rotor Construction', Ph.D. Thesis, University of St Andrews, Queens College Dundee 1966.

[42] R.W. Menzies joined the project in 1963 and made a particularly valuable contribution by completely re-engineering the digital control circuitry into a more robust Mark 2 version capable of four-path switching, which was used for all the later tests.

[43] Cruickshank, A.J.O. and Anderson, A.F.: UK Patent 1,114,561, 1968. The patent makes it clear that 'the stator member can have either salient or non-salient poles and either concentrated or distributed windings'. In this patent attention is drawn to an earlier patent by Westinghouse (UK Patent 1,010,205, 1962) in which a rotating magnetic field in a reluctance motor is produced from a single-phase supply in three separate phase windings by triggering in sequence six thyristors.

[44] This work used the method developed in Pohl's paper on 'The Theory of Pulsating Field Machines', *Proc. IEE*, Vol. 93, II, 1946, pp. 37–47, in which Pohl talks repeatedly of swinging-field machines. The main impetus for Pohl's work was the need to be able to calculate the performance of high-frequency inductor type alternators, for which standard rotating machine theory was proving unsatisfactory. The application of high-frequency power to many industrial purposes – induction furnaces, surface hardening, dielectric heating and high speed drives – had created an ever increasing demand for high-frequency alternators, motors and oscillators, converters, clutches and brakes.

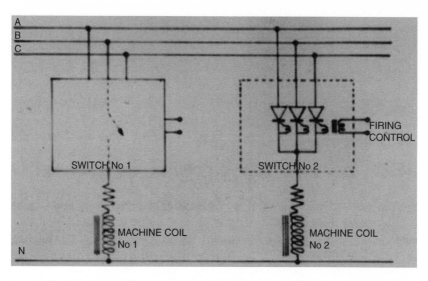

Fig. 2.20 Switching arrangement for 4 + 4 pole swinging-field reluctance motor. (Cruickshank and Anderson, UK Patent No. 1,114,561, 1968).

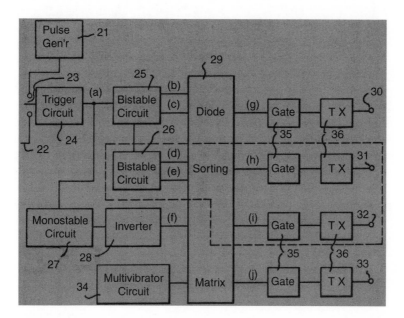

Fig. 2.21 Control circuit to drive switching circuits for swinging-field reluctance motor (two- or four-pulse arrangement). (UK Patent No. 1,114,561, 1968).

anisotropy, achieved by a combination of internal flux barriers and external shape, critically influenced performance. This led to the consideration of using axially laminated grain-oriented steel for rotors and a novel strip-wound axially laminated (ALA) rotor was developed as shown in Figure 2.13 which is also covered by UK Patent 1,114,561.

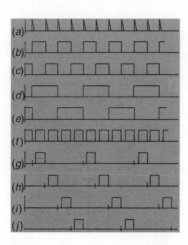

Fig. 2.22 Firing pulses that drive the switching circuits for swinging-field reluctance motor, showing four-pulse arrangement. (UK Patent No. 1,114,561, 1968).

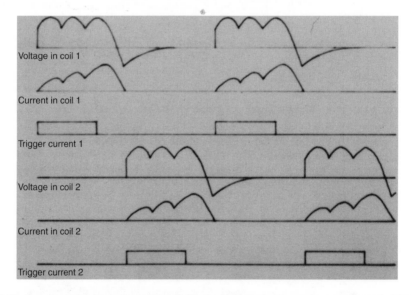

Fig. 2.23 Currents and voltages in swinging-field reluctance motor (two-pulse arrangement). (UK Patent No. 1,114,561, 1968).

Four different stator arrangements were investigated experimentally:

Stator 1: two sets of four salient poles (Figure 2.24);
Stator 2: two sets of four-pole distributed windings (Figure 2.25);
Stator 3: a conventional three-phase 4-pole stator used in a three-step stepping mode[45] (Figure 2.26);

---

[45] For every 120 degree movement of the stator energizing field clockwise the rotor advanced 60 degrees counterclockwise.

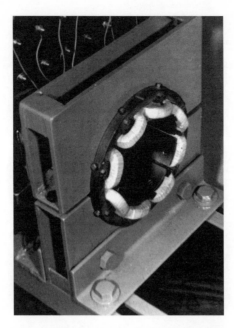

Fig. 2.24 Anderson, A.F. Ph.D. Thesis: The Thyristor Control of a Reluctance Motor of Improved Rotor Construction, University of St Andrews, Queens College Dundee, 1966: Plate 4 facing page 23.

Fig. 2.25 Stator 2: 4 + 4 distributed windings. Anderson, A.F., Ph.D. Thesis: The Thyristor Control of a Reluctance Motor of Improved Rotor Construction, University of St Andrews, Queens College Dundee, 1966: Plate 5 facing page 31.

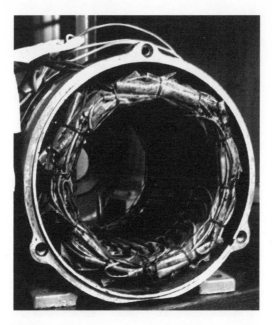

Fig. 2.26 Stator 3: three-phase 4-pole stepped in steps of 120 degrees. Anderson, A.F., Ph.D. Thesis: The Thyristor Control of a Reluctance Motor of Improved Rotor Construction, University of St Andrews, Queens College Dundee, 1966: Plate 7 facing page 35.

Fig. 2.27 Stator 4: 4 + 4 with near full pitched coils. Anderson, A.F., Ph.D. Thesis: The Thyristor Control of a Reluctance Motor of Improved Rotor Construction, University of St Andrews, Queens College Dundee, 1966: Plate 8 facing page 37.

Stator 4: two sets of four salient poles, with approximately the same m.m.f. distribution as Stator 1, achieved by using two fully pitched and concentrated coil groups displaced from one another (Figures 2.27 and 2.28).

Work on switched-field machines was not continued in Dundee after 1965 for the following reasons:

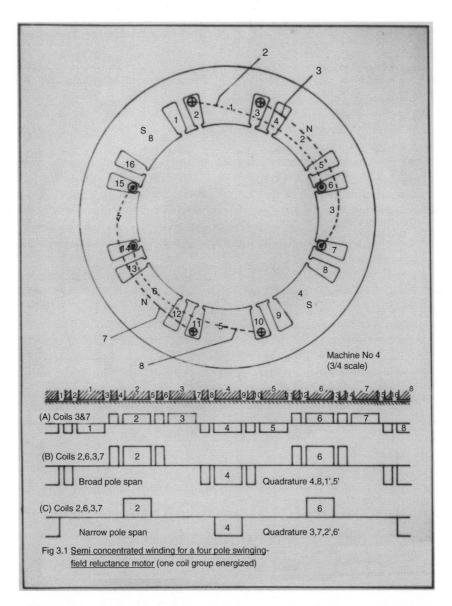

Machine No 4
(3/4 scale)

(A) Coils 3&7

(B) Coils 2,6,3,7

Broad pole span   Quadrature 4,8,1',5'

(C) Coils 2,6,3,7

Narrow pole span   Quadrature 3,7,2',6'

Fig 3.1 <u>Semi concentrated winding for a four pole swinging-field reluctance motor</u> (one coil group energized)

Fig. 2.28 Stator 4: 4 + 4 with near full pitched coils: MMF wave similar to Stator 1. Anderson, A.F., Ph.D. Thesis: The Thyristor Control of a Reluctance Motor of Improved Rotor Construction, University of St Andrews, Queens College Dundee, 1966: Figure 3.1 facing page 38.

1. The electronic hardware was still too expensive to be economic.
2. The natural commutation method used caused a beat frequency in the current pulses fed to the windings and gave very rough running at high speeds. Forced current commutation could have been adopted, but only with more complex switching circuits and hence greater cost.[46]
3. The ALA rotor looked promising as a synchronous reluctance motor rotor[47,48] and attention was therefore focused on establishing its synchronous capability.

Certain points emerge from this survey of switched reluctance motor prehistory up to 1965:

- Most switched-field machines built in the prehistorical period were doubly salient with concentrated field windings, but this was by no means always the case.
- Examples can be found of switched-field reluctance motors from the earliest times that make use of both varying self-inductance and mutual inductance. For example, Page's electromagnetic engines.
- The theory of reluctance torque derived from varying self- and mutual inductance for multi-coil reluctance type electromechanical devices was already well understood and documented and had been applied to various reluctance actuation devices.[49]
- Inductor alternator designs seem to have exploited an almost infinite variety of tooth, slot and winding distributions with the aim of achieving high-frequency outputs for relatively low rotor speeds. Some of these topologies were equally applicable to switched-field motor design.
- Pohl's work on pulsating machines, mostly of the swinging-field reluctance type, provided a sound starting point for addressing switched-field reluctance motor design that did not rely on rotating field theory.
- Pohl's work had been extended by Anderson to cover a variety of different winding and rotor configurations. This led to the conclusion that a fully pitched concentrated turn winding, if used with the appropriate kind of segmented or anisotropic rotor, might perform better than a doubly salient machine and that therefore a variety of different stator rotor topologies might be worth investigating.
- The construction and electrical connection sequences used in the Admiralty 'M' type motor had long shown that a three-phase distributed stator winding could be connected in switched reluctance mode, so that the currents in all three windings contributed to the reluctance torque.
- The axially laminated construction of the 'M' type rotor demonstrated that internal anisotropy was just as effective as external saliency in the production of reluctance torque when a reluctance motor was used in stepping mode.

---

[46] At the time power transistors were of extremely limited capability and GTO thyristors were unknown.

[47] Cruickshank, A.J.O., Anderson, A.F. and Menzies, R.W.: 'Theory and performance of reluctance motors with axially laminated anisotropic rotors', *Proc. IEE*, Vol. 118, No. 7, July 1971, pp. 887–894. This paper has an extensive bibliography on reluctance motor development up to that time.

[48] Cruickshank, A.J.O., Anderson, A.F. and Menzies, R.W.: 'Stability of Reluctance Motors from freely accelerating Torque Speed Curves', IEEE Winter Meeting, New York, NY, 30 January–4 February 1972, Paper No. T 72 049-0.

[49] Fitzgerald, A.E. and Kingsley, C.: *Electrical Machinery – The Dynamics and Statics of Electromechanical Energy Conversion*, 2nd Edition, 1952. McGraw-Hill Book Company Inc.

- The subsequent results of Cruickshank, Anderson and Menzies, with much larger axially laminated rotors, illustrated much the same point for synchronous reluctance motors.
- The possibilities of electronic commutation of electrical machine currents had already been demonstrated with mercury arc and thyristor commutators in the 1930s. The arrival of the thyristor and power transistors in the 1960s made the switched-field reluctance motor with electronic current commutation a possibility although at the time the cost was still too high to be economic.

2.5 Development since 1965[50]

2.5.1 The dawn of the modern era

The mid-1960s could be said to mark a transition from the 'historical' to the 'modern' era of switched reluctance machines, brought about by four parallel developments in the industry, all of which started in the late 1960s and early 1970s and gathered momentum firstly at a slow rate, but then at a much more rapid rate since the mid-1980s:

1. The development of the power transistor, initially the bipolar junction transistor and later the power MOSFET and the IGBT. These devices facilitated forced commutation and pulse-width modulated current regulators working at frequencies much higher than those permitted by thyristor devices.
2. The development of microprocessors and related digital integrated circuitry capable of implementing the control algorithms, which are more complex than those of d.c. and early a.c. drives (Bose, 1987).
3. The development of high-speed computers with advanced programming languages, together with powerful numerical methods for finite-element analysis and the solution of time-stepped differential equations.[51]
4. The general expansion in the use of variable-speed electric motors and motion control systems, not only in industry but in automobiles, home appliances, office products, and aerospace auxilliary systems.

2.5.2 Seminal work in the modern era

With this background many engineers in Europe, the United States and Japan began to contemplate new possibilities for electric machines and drives, as we have seen earlier in this chapter. While some of these investigations were forgotten or abandoned, some of the most notable ones produced results which can be followed to the present day.

---

[50] Section 2.5 was contributed by the editor.

[51] This period saw the publication of many sophisticated classical analyses of synchronous reluctance machines operating with variable speed, especially in relation to stability, e.g. Lawrenson and Agu (1964), Cruickshank, Anderson and Menzies (1972), Honsinger (1971) but it was also the start of a huge expansion in publication of papers on time-stepping analysis and the finite-element method, e.g. Konecny (1981). Some of the developers of switched reluctance machines (e.g. Anderson, Lawrenson & co.) were also active in the earlier development of synchronous reluctance and other a.c. drive systems; see, for example, Lawrenson and Agu (1963).

The following list is inevitably incomplete but highlights a few which are of serious interest in the technical development:

1. Cruickshank, Anderson and Menzies' development of the switched-field reluctance machine is an important example of the emergence of the 'stepped-field' concept from a seedbed of work on classical rotating-field synchronous reluctance machines (Cruickshank and Anderson, 1966a,b, 1968a,b; Anderson, 1966). Many of the key features of modern reluctance machines and their drives can be found in this work, and a number of later patents were anticipated by them, but the work was possibly just a few years too early to take advantage of the technological and commercial factors discussed in Section 2.5.1.

2. At the General Electric Company in Schenectady, NY, B.D. Bedford published two patents in 1972, describing almost all the key features of the modern switched reluctance machine and its drive (Bedford, 1972a,b).[52] This work may have been influenced by the desire to find new or better ways to exploit the power semiconductor devices that were being developed at the time at GE, but the company also manufactured a wide range of specialty motors including stepper motors and other relatives of the switched reluctance motor.

3. Lawrenson and Stephenson at the University of Leeds began work on switched reluctance motors in this period, with an already rich background of research in a.c. machines and drives, synchronous reluctance machines, stepper motors, electromagnetic field analysis, and sensors. Their programme is probably the only one which consistently and successfully developed commercial applications from the beginning, through the SR Drives company which was absorbed by Emerson in 1994. This success was achieved in the face of widespread scepticism and even hostility to the concept of the switched reluctance motor, based on fears about the noise level, the volt-ampere requirement, the complexity of the controller, and the shaft position sensing requirements. About half the commercial applications of switched reluctance motor drives since the early 1980s are attributable to this company. Lawrenson and his co-workers published seminal papers around 1980, which awakened worldwide interest in the subject (Lawrenson *et al.*, 1980).

4. Professor J.V. Byrne and his co-workers at University College, Dublin did valuable original work on machines with low phase number and drives with fewer power transistors than conventional equivalents. Byrne also contributed key material on the influence of saturation. He appears to be the first to have understood the importance of saturation in reducing the volt-ampere requirement of the drive for a given power output of the motor. Although he published this work quite widely and accessibly, it is not widely quoted and this might be partly a result of confusion about 'force doubling' which was the subject of debate in the late 1960s and throughout the 1970s. It was not until 1985 that this issue was fully resolved (Miller, 1985b). Byrne's work on 'controlled saturation' led to a number of patents, in one of which (1970) the simple mathematics of the highly saturated machine are expressed very clearly, in a way which makes the parallel with the permanent-magnet brushless motor quite clear. Byrne's research was also followed with interest by commercial companies in Europe and the United States, and a servo-motor system developed

---

[52] The 3,679,953 patent in particular is one of very few that could be described as a 'master' patent in this field.

by Inland Kollmorgen (Ireland) Ltd was introduced at the Hanover Fair in 1985,[53] although it was not put into production. See Byrne and Lacy (1976a) and Byrne and McMullin (1982).

5. The work of Vilmos Török is remarkable because it anticipates many other inventions which followed. For example, Török (1974) teaches double-stack machines, windings spanning more than one-pole, 'pre-magnetization' or bias windings, natural commutation (used with SCR-based drive circuits), and even the use of a conducting screen between the rotor poles. Török went on to develop single-phase switched reluctance motors (Török and Loreth, 1993).

6. Several other early inventions, prototypes, and research papers contributed to the momentum of development: notably Bausch and Rieke (1978, 1982), Unnewehr and Koch (1974).

2.5.3   The present status of switched reluctance technology

The present situation in switched reluctance motors and drives can perhaps be characterized as follows.

1. There is a high rate of filing patents, especially in the United States, Europe and Japan.

2. There is a very high rate of publication of academic research papers, and most conferences on electric machines include whole sessions on switched reluctance technology.

3. The number of commercial products is still very small, possibly fewer than 20. Only one or two of these have volumes of more than 100 000 per year, and many are made in quite small quantities (10–10 000 per year).

4. Many companies have built prototype switched reluctance motor drives for evaluation, but very few have realized a commercial product. Probably the main reasons are in the areas of acoustic noise, tooling cost, shaft position sensing requirement, and the complexity of the controller which must be individually tailored for each design and each application (and in some cases, even for each unit of production). The state of patents may be a discouragement to independent companies or newcomers. In spite of its positive aspects, the switched reluctance motor does not perform many functions that cannot be equally well served by more conventional technologies, most of which have benefited even more than the switched reluctance motor from the developments in power semiconductors, microelectronics, sensors, and computer methods for design and simulation. With a base of existing production and tooling, the conventional induction motor, the brushless permanent-magnet motor, and the a.c. universal motor have made such progress that it has become ever more difficult for the switched reluctance motor to compete. However, in lower-volume products or in fields where the special characteristics are extremely well suited to the application, the switched reluctance motor has good prospects. The Maytag Neptune® washing machine is an existing example (Furmanek, French and Horst, 1997).

---

[53] *Electric Drives and Controls*, June 1985, p. 45.

2.5.4   Patent activity

There has been a rapid expansion in patent activity since about 1985. Probably no single patent could claim to be the 'master patent' for switched reluctance motors or their controls, although Cruickshank and Anderson (1968a,b), Bedford's two patents (1972) and Török (1974) are interesting because of their early dates and the breadth of their teaching, and because they could be said to read on the motor and controller together as a combined system, which was uncommon before that time. The inseparability of the switched reluctance motor from its drive is one of its most fundamental characteristics, and any proposition concerning the motor without the drive, or vice versa, could be said to be incomplete. It is curious that in spite of the vigour with which the subject was pursued since about 1970, most patents fall into fairly narrow areas although some of these are clearly very important. The 'waveform patent' of Ray and Davis (1981) was important in its time in support of successful pioneering commercial developments by SR Drives Ltd, but even this patent could be said to be 'predated' by Byrne and Lacy (1976b). More recently, the invention of new motor concepts and power electronic circuits has continued at an astonishing rate considering the maturity of the underlying technology, while there is an even greater focus on sensorless control.[54] At the time of writing (April 2000) it is probably impossible to manufacture a switched reluctance motor drive having all the best known technical characteristics without using or infringing at least one of the patents written since 1985. The following paragraphs provide a very rough guide to some *examples* of significant patents; it is by no means complete and does not indicate any priority or place any weighting, neither does the absence of a patent from this section imply that it is considered unimportant. Note that there are many instances where two or more patents have similar or (apparently) identical claims.

1. *Machine geometry and windings.* A wide variety of geometric variations has resulted from the search for low noise, reduced torque ripple, improved efficiency or power density, easier manufacture, and high-speed operation, or reduced phase number. See, for example,

   (a) Compter (1985), Horst (1992, 1994), Stephenson (1993, 1996) and Byrne and Lacy (1976b) as examples of single-phase or two-phase machines.
   (b) Konecny (1986) as an example of a machine employing 'controlled saturation' to produce ripple-free torque (see also Byrne (1970)).
   (c) Hedlund (1988), Welburn (1984) and Finch *et al.*, (1984) as examples of machines with multiple teeth/pole.
   (d) Stephenson (1990) as an example of a machine with improved heat transfer properties achieved with a projection extending into the stator slot.
   (e) Weller (1990) as an example of closures fitted in the rotor slots to reduce windage loss.
   (f) Bahn (1989), Hendershot (1992), Horst (1994) Pengov (1998) as examples of machines with irregular spacing of rotor and/or stator poles.
   (g) Lipo and Liao (1994) as an example of a machine including permanent magnets.
   (h) Lipo and Liang (1993) as an example of a machine with a d.c. bias winding.

---

[54] I.e. the elimination of the shaft position sensor as used for commutation.

(i) Cruickshank and Anderson, (1968), Török, (1974), Mecrow (1992) as examples of machines with pitched windings.

2. *Sensorless control.* In some cases this is motivated by the need to reduce cost by eliminating a resolver, encoder, or Hall sensor. In other cases, such as hermetic compressors, the sensor cable presents a problem passing through the casing. Chapter 7 covers sensorless control in detail.

3. *Drive circuits and controls.* Several drive circuits have been devised to reduce the number of power semiconductor switches. Others actually increase the complexity of the drive, for example the d.c. bias winding, which is an attempt to improve the power density of the motor, or circuits which attempt to isolate different parts of the machine to improve 'fault tolerance'. Many modifications to the drive circuit are intended for acoustic noise reduction. See, for example,

(a) Byrne (1970), Miller, Steigerwald and Plunkett (1987), Van der Broeck (1988), Dhyanchand, (1991), Pollock and Williams (1991), Orthmann (1992), Sood (1992), Palaniappan *et al.* (1992), Sakano (1994) as examples of circuits with reduced number of power semiconductor switches.

(b) Miller (1985a), Hedlund (1990), Weller (1990), McHugh (1997) as examples of methods of controlling the current waveform with a 'zero-volt loop' or with pulse-width modulation of the current waveform.

(c) Ray *et al.* (1981), Bose (1987), Moreira (1992), Lovatt and Stephenson (1994), Kjaer *et al.* (1995), Kjaer *et al.* (1997), Stephenson (1997), Gribble *et al.* (1999) as examples of current waveform control and torque control.

(d) Barrass and Mecrow (1998) as an example of flux control.

(e) Sahoo *et al.* (1999) as examples of adaptive methods applied to torque control.

2.6 Summary

This chapter surveys the development of switched reluctance technology from the beginnings of electrical engineering to the present day. The history divides naturally into two eras, the first of which precedes the arrival of power transistors, microelectronics, and powerful computers in the late 1960s. The second era has been and still is characterized by a growth of technical invention, which started slowly but expanded rapidly in the 1990s. Perhaps surprisingly, the number of commercial products using switched reluctance machines is still very small, although the ones in commercial production are quite successful in their specialized fields.

3

Electromagnetic energy conversion

T.J.E. Miller
SPEED Laboratory, University of Glasgow

3.1 Definitions

A *reluctance machine* is an electric machine in which torque is produced by the tendency of its moveable part to move to a position where the inductance of the excited winding is maximized. As we have seen in Chapter 1, this definition covers both *switched* and *synchronous* reluctance machines. The switched reluctance motor has salient poles on both the rotor and the stator, and operates like a controlled stepper motor.

A primitive example is shown in Figure 3.1. This machine is denoted '2/2' because it has two stator poles and two rotor poles. The two coils wound on opposite stator poles are excited simultaneously, and generate magnetic flux as shown. There is only one phase. In the position shown, the resulting torque tends to rotate the rotor in the counterclockwise direction towards the *aligned position*, Figure 3.2. This machine can produce torque only over a limited arc of rotation, roughly corresponding to the *stator pole arc* β_s. However, it is the basic model on which the theory of torque production is based, so we will analyse this machine first, and then consider methods of starting and the extension to multiple poles and phases.

3.1.1 Aligned and unaligned positions

For the primitive reluctance machine in Figures 3.1–3.3, the aligned and unaligned positions are characterized by the properties summarized in Table 3.1.

3.1.2 Variation of inductance with rotor position

In the simple machine shown in Figures 3.1–3.3 the coil inductance L varies with rotor position θ as shown in Figure 3.4. Positive rotation is in the counterclockwise direction. Assume that the coil carries a constant current. Positive motoring torque is produced only while the inductance is increasing as the rotor approaches the aligned

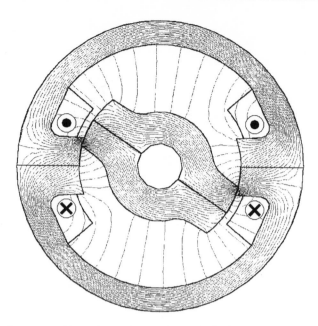

Fig. 3.1 Simple reluctance machine with one phase and two poles on both the stator and rotor.

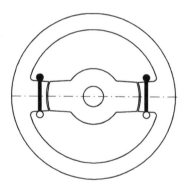

Fig. 3.2 Aligned position.

position; that is, between positions J and A. At J, the leading edge of the rotor pole is aligned with the first edge of a stator pole; at A, the rotor and stator poles are fully aligned. Thus J defines the start of overlap, A the maximum overlap, and K the end of overlap.

The torque changes direction at the aligned position. If the rotor continues past A, the attractive force between the poles produces a retarding (braking) torque. If the machine rotates with constant current in the coil, the negative and positive torque impulses cancel, and therefore the average over a complete cycle is zero. To eliminate the negative torque impulses, the current must be switched off while the poles are separating, i.e. during the intervals AK, as in Figure 3.5.

The ideal current waveform is therefore a series of pulses synchronized with the rising inductance intervals. The ideal torque waveform has the same waveform as the

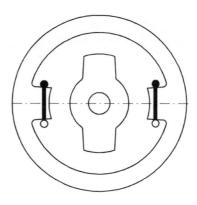

Fig. 3.3 Unaligned position.

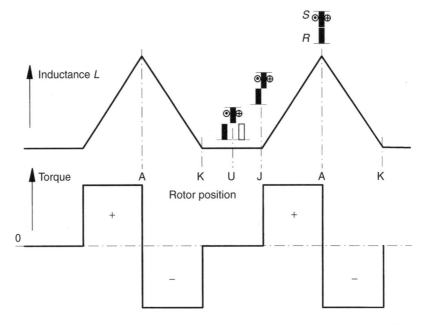

Fig. 3.4 Variation of inductance and torque with rotor position; coil current is constant. The small icons show the relative positions of the rotor and stator poles, with the rotor moving to the right. A = aligned position; U = unaligned position; J = start of overlap; K = end of overlap.

Table 3.1 Properties of the aligned and unaligned positions

Aligned	Unaligned
$\theta = 0,\ 180°$	$\theta = \pm90°$
Maximum inductance	Minimum inductance
Magnetic circuit liable to saturate	Magnetic circuit unlikely to saturate
Zero torque: stable equilibrium	Zero torque: unstable equilibrium

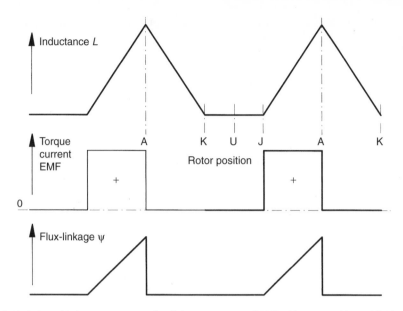

Fig. 3.5 Variation of inductance, current, flux-linkage, torque, and EMF with rotor position, with ideal pulsed unidirectional current.

current. The cycle of torque production associated with one current pulse is called a *stroke*. Evidently the production of continuous unidirectional torque requires more than one *phase*, such that the gaps in the torque waveform are filled in by currents flowing in the other phases. The numbers of phases and poles are discussed in Section 3.4.1. Normally there is one stroke per rotor pole-pitch in each phase, and the current in any phase is generally flowing for only a fraction of the rotor pole-pitch. Note that the current and inductance waveforms imply a sawtooth waveform of flux-linkage $\psi = Li$. Such a waveform is not practical because the sudden extinction of the flux and current would require an infinite negative voltage $d\psi/dt = -\infty$. Similarly the current cannot be established in step fashion unless the inductance at the beginning of the stroke (J) is zero. In practice the inductance along UJ is very small so the leading-edge di/dt can be very large, presenting a possible problem for the power semiconductors. The rectangular current waveform in Figure 3.5 can be approximated at low speed by chopping the current along JA, which has the effect of reducing the average forward applied voltage along JA to a value V_a lower than the supply voltage V_s. If there is no chopping after commutation at the end of the stroke, the reverse voltage $-V_s$ makes the current fall to zero over a very small angle of rotation.

3.2 Linear analysis of the voltage equation and torque production

Linear analysis assumes that the inductance is unaffected by the current: that is, there is no magnetic saturation. For simplicity we also ignore the effect of fringing flux around the pole corners, and assume that all the flux crosses the airgap in the radial direction.

Mutual coupling between phases is normally zero or very small, and is ignored. The voltage equation for one phase is

$$v = Ri + \frac{d\psi}{dt} = Ri + \omega_m \frac{d\psi}{d\theta}$$

$$= Ri + \omega_m \frac{d(Li)}{d\theta} = Ri + L\frac{di}{dt} + \omega_m i \frac{dL}{d\theta} \tag{3.1}$$

where v is the terminal voltage, i is the current, ψ is the flux-linkage in volt-seconds, R is the phase resistance, L is the phase inductance, θ is the rotor position, and ω_m is the angular velocity in rad/s. The last term is sometimes interpreted as a 'back-EMF' e:

$$e = \omega_m i \frac{dL}{d\theta}. \tag{3.2}$$

It is helpful to visualize the supply voltage as being dropped across the three terms in (3.1): namely, the resistance voltage drop, the $L\,di/dt$ term, and the back-EMF e. The instantaneous electrical power vi is

$$vi = Ri^2 + Li\frac{di}{dt} + \omega_m i^2 \frac{dL}{d\theta}. \tag{3.3}$$

The rate of change of magnetic stored energy at any instant is given by

$$\frac{d}{dt}\left(\frac{1}{2}Li^2\right) = \frac{1}{2}i^2\frac{dL}{dt} + Li\frac{di}{dt} = \frac{1}{2}i^2\omega_m\frac{dL}{d\theta} + Li\frac{di}{dt}. \tag{3.4}$$

According to the law of conservation of energy, the mechanical power conversion $p = \omega_m T_e$ is what is left after the resistive loss Ri^2 and the rate of change of magnetic stored energy are subtracted from the power input vi, T_e being the instantaneous *electromagnetic torque*. Thus from (3.2) and (3.3), writing $T_e = p/\omega_m = vi - Ri^2 - d(1/2Li^2)/dt$, we get

$$T_e = \frac{1}{2}i^2\frac{dL}{d\theta}. \tag{3.5}$$

Note that $dL/d\theta$ is the slope of the inductance graph in Figure 3.5.

Drive circuit – unidirectional current, bidirectional voltage
Equation (3.5) says that the torque does not depend on the direction of the current, since i^2 is always positive. However, the voltage must be reversed at the end of each stroke, to return the flux-linkage to zero. By Faraday's law, this requires a negative voltage applied to the coil, to ensure that $d\psi/dt < 0$. Figure 3.6 shows the *half-bridge phaseleg* circuit that accomplishes this requirement. When Q1 and Q2 are both on, the voltage across the motor windings is $v = V_s$, the supply voltage. When Q1 and Q2 are both off, $v = -V_s$ while the current freewheels through D1 and D2.

In order to reduce ψ and i to zero as quickly as possible, as in Figure 3.5, it seems that the reverse voltage must be much larger than the forward voltage, otherwise the flux-linkage will persist beyond the aligned position, producing an unwanted negative pulse of torque. At low speeds this is achievable by chopping the forward voltage, reducing its effective value in proportion to the duty-cycle of the chopping.

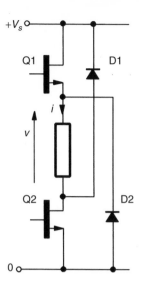

Fig. 3.6 Single phaseleg circuit.

The circuit of Figure 3.6 is capable of operating the machine as a motor or as a generator, since the electrical power vi can be positive or negative. If the average power is negative (i.e. generating), the energy supplied during transistor conduction in one stroke must be less than the energy recovered during freewheeling. The transistor conduction period (with positive applied voltage) is still necessary to establish the flux, which is built up from zero and returned to zero each stroke. The voltage–time integrals during transistor conduction and freewheeling must be approximately equal (aside from the effect of resistance voltage-drop), regardless of whether the machine is motoring or generating. From the control point of view, the main difference between motoring and generating is the phasing of the conduction pulse relative to the rotor position. From Figure 3.4 it is evident that generating current pulses must coincide with AK, just as motoring pulses coincide with JA.

Several other circuits are possible besides the one in Figure 3.6, including some which use only one transistor per phase, and employ various means to produce the 'suppression voltage' (i.e. the reverse voltage) needed to de-flux the windings at the end of each stroke. For example, the circuit in Figure 3.7a uses separate voltage sources for 'fluxing' and 'de-fluxing'. The circuit in Figure 3.7b goes one stage further by separating the windings into two isolated circuits with a common magnetic circuit but no galvanic coupling. The two parts of the winding could be bifilar-wound or they could have completely different numbers of turns and wire sizes. Unfortunately the leakage inductances of the two parts of the winding are usually quite large, even with a bifilar winding, and this leads to problems with transistor overvoltage. Consequently the circuit of Figure 3.7b is rarely used, although it is common in stepper-motor drives.

Additional phases simply use additional drive circuits of the same form as the first phase, usually with a common voltage source. Chapter 8 describes a circuit for a four-phase motor which uses a common chopping transistor between two phases, so that only six power transistors are required for four phases. The control of the drive is discussed in more detail in Chapter 5 and subsequent chapters.

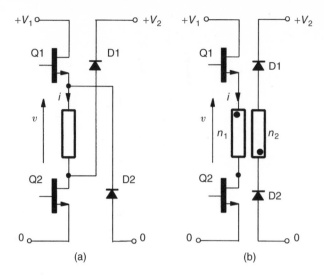

Fig. 3.7 Alternative single-phaseleg circuits.

Limitations of the ideal linear model

At the aligned position A in Figure 3.5, the current must be switched off quickly to avoid the production of negative torque after the poles have passed the aligned position. The magnetic stored energy $\frac{1}{2}L_a i^2$ must be returned to the supply. In a nonsaturating reluctance machine of this type, the magnetic energy stored at A is large, because both L and i are at their maximum values. We can get further insight from an energy audit taken over one stroke as the rotor moves from J to A. The process is shown in the energy conversion diagram, Figure 3.8, which plots flux-linkage against current. The slope of OJ is the inductance at the unaligned position, L_u, and the slope of OA is the inductance at the aligned position, L_a. At intermediate positions the inductance is represented by a line of intermediate slope between L_u and L_a. At the J position (start of overlap), if fringing is neglected (as in the idealized inductance variation in Figure 3.4), $L_J = L_u$. The complete stroke is represented by the locus OJAO. In motoring operation it is traversed in the counterclockwise direction, and in generating operation in the clockwise direction.

Although the current in Figure 3.5 has a step from 0 to a maximum value i_m at the position J, in Figure 3.8 this step is along OJ and energy $OJC = \frac{1}{2}L_u i_m^2$ must be supplied to the magnetic circuit as the current increases from 0 to i_m. The step cannot be accomplished in zero time, since that would require infinite voltage, but if the angular velocity is low, the angle of rotation along OJ is small. Along JA the electromechanical energy conversion is W or OJA, given by

$$W = \text{OJA} = T_e \Delta\theta = \frac{1}{2}i_m^2 \frac{dL}{d\theta}\Delta\theta = \frac{1}{2}i_m^2 \Delta L. \tag{3.6}$$

Along JA there is a back-EMF e which absorbs energy equal to the area ABCJ:

$$ei\,\Delta t = \text{ABCJ} = i_m^2 \omega_m \frac{dL}{d\theta} \times \frac{\Delta\theta}{\omega_m} = i_m^2 \Delta L \tag{3.7}$$

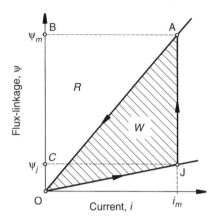

Fig. 3.8 Linear energy conversion diagram.

where $\Delta t = \Delta\theta/\omega_m$ is the time taken to rotate through the interval $\Delta\theta = $ JA, and $\Delta L = L_a - L_u$ is the change in L. The total energy supplied is the sum of areas OCJ and ABCJ, i.e. area $S = W + R = $ OJAB, and

$$S = i_m^2 \Delta L + \frac{1}{2} i_m^2 L_u. \tag{3.8}$$

We can now define the *energy ratio* as

$$Q = \frac{\text{energy converted}}{\text{energy supplied}} = \frac{W}{S} = \frac{W}{W + R} \tag{3.9}$$

and if we write $\lambda = L_a/L_u$ we can substitute for W and S from (3.6) and (3.8) to give

$$Q = \frac{\lambda - 1}{2\lambda - 1}. \tag{3.10}$$

In this type of nonsaturating motor, less than half the energy S supplied by the drive is converted into mechanical work in each stroke (even neglecting losses). During the 'working' part of the stroke JA, the energy is partitioned equally between mechanical work and stored field energy; this is evident from the ratio of (3.6) and (3.7), or OJA/ABCJ. The energy ratio would have a value of 0.5 but for the 'overhead' of stored field energy OJC which must be built up before the torque zone JA. This makes $Q < 0.5$. The inverse of the energy ratio is the converter volt-ampere-seconds per joule of energy conversion $C = S/W = 1 + R/W$, an important quantity in understanding the basic requirement for 'silicon' in the converter.

The stored field energy reaches a maximum value R at A, and must be returned to the supply at the end of the stroke by commutating the current into the diodes, so that the voltage reverses and forces the flux-linkage to fall to zero. In the ideal locus this fall is along AO. However, this is not possible with finite supply voltage. If the current is chopped with a duty-cycle d along JA, the average forward applied voltage along JA is $d \times V_s$. In a circuit of the form of Figure 3.6, the reverse voltage after commutation is $-V_s$. By integrating Faraday's law the rise and fall periods of flux-linkage can be shown to be in the ratio

$$\frac{t_f}{t_r} = d. \tag{3.11}$$

This shows that 'instant' suppression of the flux is effectively achieved at very low speed when d is small. But at higher speed there is no chopping: the forward and reverse voltages are equal in magnitude, and $d = 1$, so $t_f = t_r$ and the flux-linkage waveform is triangular. The time taken along OA is the same as the time taken along OJA. The waveforms of flux-linkage and current corresponding to Figure 3.8 are shown in Figure 3.9, and show a tail in the current waveform extending past the aligned position so that some negative torque must be produced.

Figures 3.8 and 3.9 represent an important operating condition where the back-EMF is just sufficient to maintain a flat-topped current waveform. With full voltage applied, the speed at which this occurs is called the *base speed*. The current i_m could be called the base current. (It is not necessarily the *rated* current, because that depends on the cooling arrangements.)

In practical terms the nonsaturating switched reluctance motor makes poor use of the power semiconductors, because they have to supply more than 2 J of energy in order to get 1 J of mechanical work, and they must also provide a means to recover the unconverted energy at the end of the stroke. The d.c. link filter capacitance is directly related to the value of R, which is a large fraction of the energy conversion W: in fact

$$R = W \times \frac{1 - Q}{Q}. \tag{3.12}$$

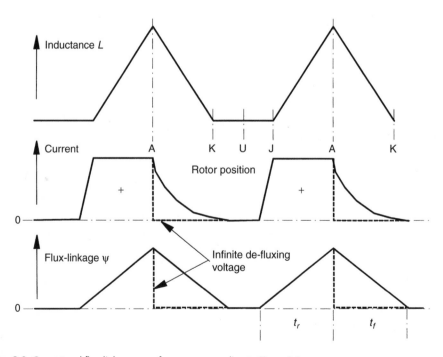

Fig. 3.9 Current and flux-linkage waveforms corresponding to Figure 3.8.

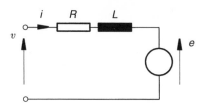

Fig. 3.10 Equivalent circuit.

For example, if $\lambda = 8$, $Q = 0.467$, $C = 2.14$, and $R = 1.14W$. This implies that for zero ripple voltage at the d.c. link, the filter capacitance must be large enough to absorb R joules with negligible change of voltage. This requirement may be reduced by overlapping charge/discharge requirements of adjacent phases, but still it is a serious consideration.

Equations (3.1) and (3.2) imply the existence of an equivalent circuit of the form shown in Figure 3.10, in which there is a back-EMF. $e = \omega_m i \, dL/d\theta$.

Unfortunately e is not an independent parameter, but depends on the current. In a normal equivalent circuit we interpret the product ei as the electromechanical power conversion $\omega_m T_e$, implying that $T_e = i^2 \, dL/d\theta$. However, (3.5) states that the torque is only $\frac{1}{2} i^2 \, dL/d\theta$. Of the power ei, *only half* is converted into mechanical power during the 'working' part of the stroke JA. The other half is being stored as magnetic field energy in the increasing inductance. With L also varying, the equivalent circuit is misleading and cannot be interpreted in the same way as it can, for example, with permanent-magnet motors. This means that the simulation of switched reluctance machines and their drives requires the direct solution of (3.1) and (3.5), even when saturation is ignored. An elegant and thorough solution for the nonsaturating motor was presented by Ray and Davis (1979). Usually saturation *cannot* be ignored and the full nonlinear equations must be solved.

3.3   Nonlinear analysis of torque production

We have already seen that the nonsaturating (i.e. magnetically linear) switched reluctance machine has a low energy ratio and makes poor utilization of the drive. Practical switched reluctance machines are more effective but they are far from being magnetically linear. To understand the electromechanical energy conversion properly, we need a nonlinear analysis that takes account of the saturation of the magnetic circuit. One such analysis is based on *magnetization curves*. A magnetization curve is a curve of flux-linkage ψ versus current i at a particular rotor position, Figure 3.11.

We also need to define the stored magnetic energy W_f and the *co-energy* W_c graphically, as in Figure 3.12. Mathematically,

$$W_f = \int i \, d\psi;$$

$$W_c = \int \psi \, di. \tag{3.13}$$

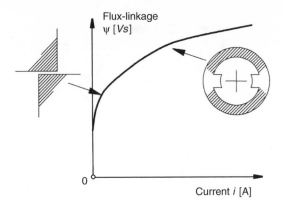

Fig. 3.11 A magnetization curve at one rotor position.

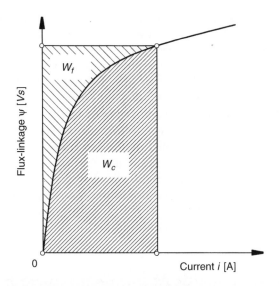

Fig. 3.12 Definition of stored field energy W_f and co-energy W_c.

In a magnetically linear device with no saturation, the magnetization curve is straight, and $W_f = W_c$. The effect of saturation is to make $W_f < W_c$. In switched reluctance machines with magnetic circuits similar to the one in Figure 3.1, saturation of a typical magnetization curve occurs in two stages. When the overlap between rotor and stator pole corners is quite small, the concentration of flux saturates the pole corners, even at quite low current. When the overlapping poles are closer to the aligned position, the yokes saturate at high current, tending to limit the maximum flux-linkage. Magnetization curves near the aligned position may appear 'double-jointed' as in Figure 3.15, especially if the airgap is small and the curve is plotted to extreme values of flux-density.

In a displacement $\Delta\theta$ or AB at constant current (Figure 3.13), the energy exchanged with the supply is

$$\Delta W_e = \int ei \, dt = \int i \frac{d\psi}{dt} dt = \int i \, d\psi = ABCD \tag{3.14}$$

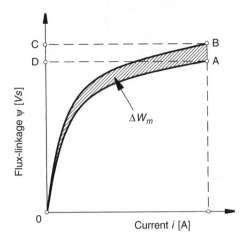

Fig. 3.13 Determination of electromagnetic torque.

and the change in magnetic stored energy is

$$\Delta W_f = \text{OBC} - \text{OAD}. \tag{3.15}$$

The mechanical work done must be

$$\Delta W_m = \Delta W_e - \Delta W_f$$
$$= \text{ABCD} - (\text{OBC} - \text{OAD}) \tag{3.16}$$
$$= \text{OABCD} - \text{OBC}$$
$$= \text{OAB}$$

and this is equated to $T_e\Delta\theta$, so that, in the limit, when $\Delta\theta \to 0$,

$$T_e = \left[\frac{\partial W_c}{\partial \theta}\right]_{i=const}. \tag{3.17}$$

Ideal cases
In a motor with no saturation the magnetization curves would be straight lines, Figure 3.14(a). At any position, $W_f = W_c = \frac{1}{2}L(\theta)i^2$, and in this case (3.17) reduces to $T_e = \frac{1}{2}i^2\,dL/d\theta$, which we saw earlier. In a motor with a very small airgap, and especially if the steel has a 'square' B/H curve, the magnetization curves approximate to parallel straight lines with a shallow slope, Figure 3.14(b). In this case the stored field energy is small and $W_c \sim \psi i$, so $T_e = i\,d\psi/d\theta = ei$. This condition approximates to the permanent-magnet motor which has an EMF that is independent of the current. The energy ratio is 1, and the utilization of the power semiconductors in the drive is high. Practical reluctance motors, especially highly rated ones, are often designed to try to approximate this ideal condition.

Average torque
So far we have seen that the torque is produced in impulses as the rotor rotates, and we have determined the *instantaneous* torque T_e in terms of the rate of change of

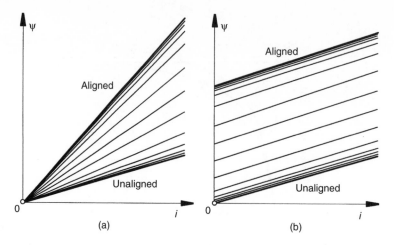

Fig. 3.14 Ideal magnetization curves with no saturation.

coenergy at constant current. The *average* torque could be calculated by integrating T_e over one cycle (i.e. one rotor pole-pitch $\tau = 2\pi/N_r$, where N_r is the number of rotor poles), and dividing by τ. However, it is difficult in practice to calculate T_e accurately, and it is better to calculate the average torque from the enclosed area W in the energy-conversion diagram, Figure 3.15.

In one cycle of operation the maximum possible energy conversion at a current I is the area W enclosed between the *unaligned* magnetization curve U, the *aligned* magnetization curve A, and the vertical line UA at the current I. One cycle of operation, i.e. one execution of this loop, is called a *stroke*. If S is the number of strokes per revolution, the average electromagnetic torque in Nm is

$$T_{e[avg]} = \frac{SW}{2\pi}. \tag{3.18}$$

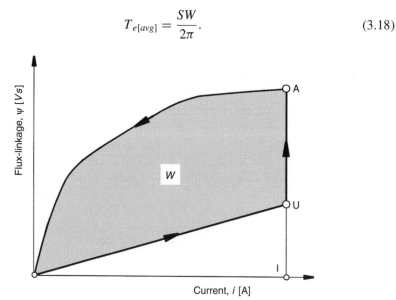

Fig. 3.15 Calculation of average torque.

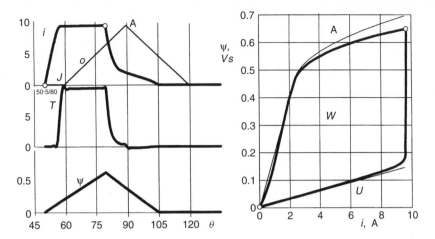

Fig. 3.16 Current, torque, and flux-linkage waveforms with a naturally determined flat-topped current waveform. i = phase current; T = phase electromagnetic torque; o = per-unit overlap between active stator and rotor poles; ψ = phase flux-linkage; W = energy conversion loop area; U = unaligned; A = aligned.

The drive must switch the current on and off at the correct rotor angles to cause the operating point to follow this loop as closely as possible. Along UA, the current is generally regulated by chopping, but at high speed this may not be achievable and the actual loop may be smaller than the maximum loop; see Figure 5.3.

An example

Figure 3.16 shows an example of a motor operating with a naturally flat-topped current waveform, which is obtained when the back-EMF is approximately equal to the d.c. supply voltage. This motor is analysed in more detail in Chapter 5. The speed in this example is 1015 rpm.

3.4 Continuous torque production

The motor in Figure 3.1 is useful for developing the analysis of torque production, and although it can maintain a nonzero average torque when rotating in either direction, this torque is discontinuous, which means that continuous rotation depends on the momentum or flywheel effect of the rotating inertia (of motor + load); or, in the case of a generator, it depends on the prime mover. Moreover it cannot self-start from every rotor position. For example, at the unaligned and aligned positions the torque is zero. Unidirectional torque can be produced only over a limited angle where the overlap angle λ between the rotor and stator poles is varying. To provide continuous unidirectional torque, with self-starting capability from any rotor position, the motor is generally provided with additional *phases* which lead to a 'multiplicity' of stator and rotor poles, as in Figure 3.17.

The number of strokes per revolution is related to the number of rotor poles N_r and the number of phases m, and in general

$$S = mN_r. \tag{3.19}$$

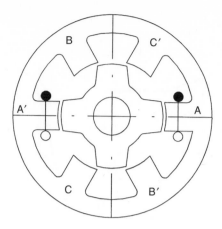

Fig. 3.17 Three-phase 6/4 switched reluctance motor.

This can be substituted in (3.18) to give the average torque including all m phases, provided that S is the same for all of them. The motor in Figure 3.17 has $m = 3$ and $N_r = 4$, so $S = 12$. The *stroke angle* is $\varepsilon = 360/12 = 30°$. The three phases are labelled AA′, BB′ and CC′, and the ideal current/torque pulses are shown in Figure 3.18. The resultant torque is ideally constant and covers 360° of rotation. In practice, of course, the torque waveform is usually far from the ideal constant-torque waveform in Figure 3.18, and its computation requires a fairly sophisticated simulation of the transient electromagnetic behaviour throughout one stroke.

Magnetic frequency

The fundamental frequency f_1 of the current in each phase is evidently equal to the rotor pole passing frequency, i.e.

$$f_1 = \frac{\text{rpm}}{60} \times N_r \text{ Hz.} \tag{3.20}$$

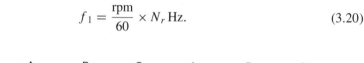

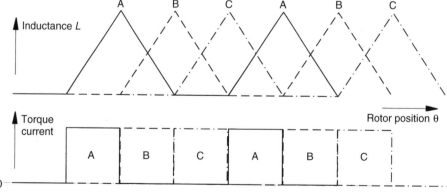

Fig. 3.18 Waveforms in three-phase 6/4 switched reluctance machine.

The number of strokes per second is given by

$$f = mf_1 \quad \text{Hz.} \tag{3.21}$$

This frequency and all its harmonics appear in the flux waveforms in various parts of the magnetic circuit.

3.4.1   Stator/rotor pole numbers

To provide a structure for ordering the numbers of stator and rotor poles, we can start by defining a *regular* switched reluctance motor as one in which the rotor and stator poles are symmetrical about their centrelines and equally spaced around the rotor and stator respectively. An *irregular* motor is one which is not regular. This chapter, indeed the whole book, is more concerned with regular machines, since they usually have the most sophisticated power electronic control requirements; and therefore it focuses on machines with $m = 3$ or 4 phases. Machines with $m = 1$ or 2 are usually irregular.

The *absolute torque zone* τ_a is defined as the angle through which one phase can produce nonzero torque in one direction. In a regular motor with N_r rotor poles, the maximum torque zone is $\tau_{a(max)} = \pi/N_r$. The *effective torque zone* τ_e is the angle through which one phase can produce *useful* torque comparable to the rated torque. The effective torque zone is comparable to the lesser pole-arc of two overlapping poles. For example, in Figure 3.17 the effective torque zone is equal to the stator pole-arc: $\tau_e = \beta_s = 30°$.

The *stroke angle* ε is given by $2\pi/(\text{strokes/rev})$ or

$$\varepsilon = \frac{2\pi}{S} = \frac{2\pi}{mN_r}. \tag{3.22}$$

The *absolute overlap ratio* ρ_a is defined as the ratio of the absolute torque zone to the stroke angle: evidently this is equal to $m/2$. A value of at least 1 is necessary if the *regular* motor is to be capable of producing torque at all rotor positions. In practice a value of 1 is not sufficient, because one phase can never provide rated torque throughout the absolute torque zone in both directions. The *effective overlap ratio* ρ_e is defined as the ratio of the effective torque zone to the stroke angle, $\rho_e = \tau_e/\varepsilon$. For regular motors with $\beta_s < \beta_r$ this is approximately equal to β_s/ε. For example, in Figure 3.17 the effective overlap ratio is $30°/30° = 1$. Note that $\rho_e < \rho_a$. A value of ρ_e of at least 1 is necessary to achieve good starting torque from all rotor positions with only one phase conducting, and it is also a necessary (but not sufficient) condition for avoiding torque dips.

Three-phase regular motors
With $m = 3$, $\rho_a = 1.5$ and ρ_e can have values of 1 or more, so *regular* three-phase motors can be made for four-quadrant operation. In the 6/4 motor in Figure 3.17, forward rotation corresponds to negative phase sequence. This is characteristic of *vernier* motors, in which the rotor pole-pitch is less than π/m. The three-phase 6/4 motor has $S = mN_r = 12$ strokes/rev, with a stroke angle $\varepsilon = 30°$, giving $\rho_e = \beta_s/\varepsilon = 30/30 = 1$.

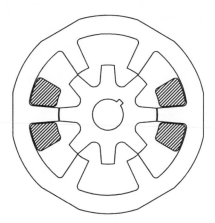

Fig. 3.19 Three-phase 6/8 motor. Each phase has two coils on opposite poles.

With regular vernier motors there is always the choice of having either $N_r = N_s - 2$, as in the 6/4; or $N_r = N_s + 2$, which gives the 6/8 motor shown in Figure 3.19; it has $S = 24$ strokes/rev and $\varepsilon = 15°$ and is similar to Konecny's motor used in the Hewlett-Packard Draftmaster plotter. The advantage of the larger N_r is a smaller stroke angle, leading possibly to a lower torque ripple; but inevitably the price paid is a lower inductance ratio which may increase the controller volt-amperes and decrease the specific output. The stator pole arc has to be reduced below that of the 6/4 motor and this decreases the aligned inductance, the inductance ratio, and the maximum flux-linkage (although it increases the slot area). The consequent reduction in available conversion energy tends to offset the increase in the number of strokes/rev, and the core losses may be higher than those of the 6/4 motor because of the higher switching frequency.

The 12/8 three-phase motor is effectively a 6/4 with a 'multiplicity' of two. It has $S = 24$ strokes/rev, with a stroke angle $\varepsilon = 15°$ and $\rho_a = 1.5$. In Figure 3.20, $\rho_e = 15/15 = 1$, the same as for the 6/4 motor discussed earlier. A high inductance ratio can be maintained and the end-windings are short: this minimizes the copper losses, shortens the frame, and decreases the unaligned inductance. Moreover, the

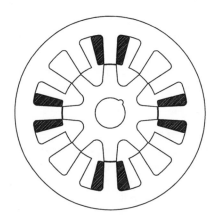

Fig. 3.20 Three-phase 12/8 motor. Each phase has four coils, and the magnetic flux-pattern is 4-pole.

magnetic field in this machine has short flux-paths because of its four-pole magnetic field configuration, unlike the two-pole configuration in the 6/4 (or the 8/6; see below), and the four-pole magnetic circuit helps to minimize acoustic noise (see Chapter 4). Although the MMF per pole is reduced along with the slot area, the effects of long flux-paths through the stator yoke are alleviated. The 12/8 is possibly the most popular configuration for three-phase machines.

Four-phase regular motors

The four-phase regular 8/6 motor shown in Figure 1.1 has 24 strokes/rev and a stroke angle of 15°, giving $\rho_a = 2$. With $\beta_s = 21°$, $\rho_e = 1.33$, which is sufficient to ensure starting torque from any rotor position, and it implies that there will be no problem with torque dips. However, it is generally impossible to achieve the same flux-density waveform in every section of the stator yoke, because of the polarities of the stator poles (NNNNSSSS, NNSSNNSS, or NSNSNSNS). This configuration was one of the first to be produced commercially.[1]

With $N_s = N_r + 2 = 10$, $S = 32$ strokes/rev and $\varepsilon = 11.25°$. The inductance ratio is inevitably lower than in the 8/6, and the poles are narrower, while the clearance between pole-corners in the unaligned position is smaller, increasing the unaligned inductance. This motor is probably on the borderline where these effects cancel each other out; with higher pole-numbers, the loss of inductance ratio and energy-conversion area tends to dominate the gain in strokes/rev. For this reason, higher pole-numbers are not considered here (but see Chapter 11).

Table 3.2 Examples of valid stator/rotor pole number combinations (preferred combinations are **not** shaded)

m	N_s	N_r	N_{wkPP}	$\varepsilon°$	S	m	N_s	N_r	N_{wkPP}	$\varepsilon°$	S
2	4	2	1	90.00	4	3	6	2	1	60.00	6
2	8	4	2	45.00	8	3	6	4	1	30.00	12
2	4	6	1	30.00	12	3	6	8	1	15.00	24
2	8	12	2	15.00	24	3	12	8	2	15.00	24
2	12	18	3	10.00	36	3	18	12	3	10.00	36
2	16	24	4	7.50	48	3	24	16	4	7.50	48

m	N_s	N_r	N_{wkPP}	$\varepsilon°$	S	m	N_s	N_r	N_{wkPP}	$\varepsilon°$	S
4	8	6	1	15.00	24	5	10	4	1	20.00	18
4	16	12	2	7.50	48	5	10	6	1	12.00	30
4	24	18	3	5.00	72	5	10	8	1	9.00	40
4	32	24	4	3.75	96	5	10	12	1	6.00	60
4	8	10	1	9.00	40						

m	N_s	N_r	N_{wkPP}	$\varepsilon°$	S	m	N_s	N_r	N_{wkPP}	$\varepsilon°$	S
6	12	10	1	6.00	10	7	14	10	1	5.14	70
6	24	20	2	3.00	120	7	14	12	1	4.29	84
6	12	14	1	4.29	84						

---

[1] I.e. the well-known OULTON motor introduced in 1983 by Tasc Drives Ltd, Lowestoft, England.

Table 3.2 gives some examples of stator/rotor pole-number combinations for motors with up to $m = 7$ phases. The parameter N_{wkPP} is the number of working pole-pairs: that is, the number of pole-pairs in the basic magnetic circuit. For example, the four-phase 8/6 has $N_{wkPP} = 1$ (a two-pole flux pattern), while the three-phase 12/8 has $N_{wkPP} = 2$ (a 4-pole flux pattern). The unshaded boxes in Table 3.2 are probably the best choices, the others having too many poles to achieve a satisfactory inductance ratio or too high a magnetic frequency.

3.5   Energy conversion analysis of the saturated machine

3.5.1   Energy ratio and converter volt-ampere requirement

Figure 3.21 shows a model of the energy conversion process in which the unaligned magnetization curve is assumed to be straight, while the aligned curve is composed of two sections, a straight line OS and a parabola SA. For a given peak current i_m the energy conversion capability W is completely defined by the three points U, S, A. It is shown in Miller and McGilp (1990) that the parabola section SA is represented by

$$(\psi - \psi_{s0})^2 = 4a(i - i_{s0}) \tag{3.23}$$

where $i_{s0} = i_s - a/L_{au}^2$, $\psi_{s0} = \psi_s - 2a/L_{au}$, $a = \psi_{ms}^2/4(i_{ms} - \psi_{ms}/L_{au})$, $\psi_{ms} = \psi_m - \psi_s$, and $i_{ms} = i_m - i_s$. These relationships ensure that the first derivatives of the segments OS and SA are equal at S. The area R can be calculated by direct integration:

$$R = \psi_{ms} \left[\frac{\psi_m^2 + \psi_s(\psi_m + \psi_s)}{12a} + \psi_{s0} \frac{\psi_{s0} - (\psi_m + \psi_s)}{4a} + i_{s0} \right] + \frac{1}{2} L_{au} i_s^2 \tag{3.24}$$

and then the energy-conversion area W can be obtained as

$$W = i_m \psi_m - R - \tfrac{1}{2} L_u i_m^2. \tag{3.25}$$

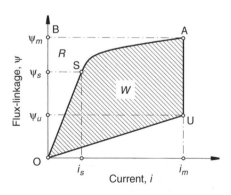

Fig. 3.21 Nonlinear energy-conversion analysis, with aligned magnetization curve represented by a straight line OS and a parabola SA.

The energy ratio Q is determined as $W/(W + R)$ together with its inverse $C = 1 + R/W$. As an example, Figure 3.21 is for a machine with an unsaturated inductance ratio $\lambda = L_{au}/L_u = 8$, and is scaled to the following approximate values: $\psi_m = 0.7$ V-s, $\psi_s = 0.5$ V-s, $\psi_u = 0.25$ V-s, $i_m = 80$ A, and $i_s = 20$ A. The resulting values are approximately $W = 27.7$ J, $R = 13.3$ J, $Q = 0.67$, and $C = 1.5$. The energy ratio is about 45% greater than in the linear nonsaturating machine ($Q = 0.467$), and the converter volt-ampere-second requirement of the nonsaturating machine is about 45% higher.

3.5.2   Estimation of the commutation angle

Figure 3.22 shows a model of the energy conversion process in which the unaligned magnetization curve is again assumed to be straight, while the aligned curve is composed of two straight lines OS and SA. For a given peak current i_m the energy conversion capability W is again completely defined by the three points U, S, A. This model closely resembles the one which J.V. Byrne (1970) used in a patent on controlled saturation (see also Byrne and Lacy (1976a)). The saturation effect is characterized by the ratio

$$\sigma = \frac{\psi_m/i_m}{\psi_s/i_s} \tag{3.26}$$

which is effectively the ratio of the saturated to the unsaturated inductance in the aligned position. At the base speed the current waveform is naturally flat-topped with peak value i_m, because the back-EMF e of the motor equals the supply voltage (resistance is neglected). If the angular rotation JA is assumed equal to the stator pole-arc β_s, then it can be shown (Miller, 1985b) that $e = V_s$ when

$$\frac{V_s\beta_s}{\omega_m} = i_m L_u(\lambda\sigma - 1) = \Delta\psi_m. \tag{3.27}$$

Commutation is at point C such that the change of flux-linkage between J and C is $c\Delta\psi_m$, where $c \le 1$. After commutation the current continues to flow throughout the angle $(1 - c)\beta_s$ and for an undetermined interval $k\beta_s$ thereafter. Since the peak

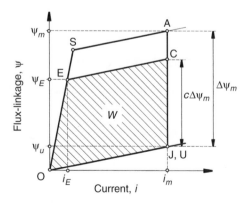

Fig. 3.22 Nonlinear energy-conversion analysis, with aligned magnetization curve represented by two straight lines.

flux-linkage ψ_c is $c\Delta\psi_m + L_u i$, we can write

$$\psi_c = \frac{V_s \beta_s}{\omega_m}(1 - c + k) = c\Delta\psi_m + L_u i_m \tag{3.28}$$

which describes the de-fluxing interval with $-V_s$ applied via the diodes. (3.27) and (3.28) can be used to solve for c:

$$c = \frac{(1+k)(\lambda\sigma - 1) - 1}{2(\lambda\sigma - 1)}. \tag{3.29}$$

When the rotor reaches the aligned position the flux-linkage is

$$\psi_E = \frac{k}{1 - c + k}\psi_c \tag{3.30}$$

and it follows that the current at the aligned position is

$$i_E = \frac{\psi_E}{L_{au}}. \tag{3.31}$$

The example in Section 3.5.1 has $\sigma = 0.350$ and $\lambda = 8$. We can consider two extreme cases:

1. *Commutation at the aligned position* – In this case $c = 1$ and from (3.29) $k = 1.556$, so the current continues to flow for 1.556 times the angle β_s after the aligned position. The flux-linkage at the aligned position is ψ_m and the current is $i_E = \psi_m/L_{au} = 40\psi_m = 28.0$ A.
2. *Commutation such that the current extinguishes at the aligned position* – In this case $k = 0$ and from (3.29), $c = 0.222$, and $\psi_E = i_E = 0$. Although all negative torque is completely eliminated, the energy conversion with $c = 0.222$ is far below the capability of the machine.

In practice the commutation angle is selected to maximize the torque per ampere, and falls between these extremes, with $k > 0$ and some negative torque. In the example shown in Figure 3.16, the negative torque is very small and $c = 2/3$, giving $k = 0.89$ and $\psi_E = 0.727\psi_c$. Although resistance accelerates the flux suppression, it can be seen that the current tail in the negative torque zone has little impact on the overall energy conversion.

3.5.3   Basic torque/speed characteristic

An interesting, simplified analysis of the torque/speed characteristic was given by Byrne and McMullin (1982), in which they derived a formula for the speed range at constant power. If ω_p is the maximum speed at which the power can be developed equal to the maximum power at base speed ω_b, then in terms of the parameters used in this chapter,

$$n = \frac{\omega_p}{\omega_b} = \frac{1}{2}\left[\frac{\theta_p}{\theta_b}\right]^2 \frac{(\lambda\sigma)^2}{\lambda\sigma - 1} \tag{3.32}$$

where θ_p and θ_b are the dwell angles (transistor conduction angles) at the speeds ω_p and ω_b respectively. For the example motor $\lambda\sigma = 8 \times 0.35 = 2.8$ and if we assume

$\theta_p = 1.5\theta_b$, we get $n = 4.9$. This is probably optimistic but it shows the importance of phase advance and the saturated inductance ratio $\lambda\sigma$. Chapter 5 shows in more detail how the torque/speed characteristic is developed. A typical speed range at constant power is probably nearer 3.

The parameter c in Figure 3.22 effectively controls the energy conversion loop area and the average torque. At low speeds it is possible to work with $c = 1$, so that the entire available energy conversion area between the aligned and unaligned curves, bounded on the right by the peak current, can be used. Miller (1985b) gives this average electromagnetic torque the equation

$$T_a = i_m^2 L_u(\lambda\sigma - 1)\frac{mN_r}{4\pi}c(2 - c/s)\qquad(3.33)$$

where $s = (\lambda - 1)/(\lambda\sigma - 1)$. At the base speed c has a value $c_b < 1$ and from (3.33) the ratio of the torque at zero speed to the torque at base speed is derived as

$$\frac{T_0}{T_b} = \frac{2s - 1}{c_b(2s - c_b)}.\qquad(3.34)$$

In the example motor $c_b = 2/3$ and $s = (8 - 1)/(8 \times 0.35 - 1) = 3.89$, and $T_0/T_b = 1.43$. The motor can evidently produce some 43% more torque at standstill than at the base speed, for the same peak current. (The r.m.s. current must be increased because the dwell is greater.) With the same peak current, the peak torque is the same in both cases and this implies that the torque waveform must be peakier at the base speed than at standstill. In the same reference Miller (1985b) goes on to compare the volt-ampere requirements of the switched reluctance motor with those of a comparable high-efficiency induction motor, and concludes that the switched reluctance motor requires 14% more volt-amperes based on peak current, or 20% more based on r.m.s. current. Such comparisons are, however, extremely difficult to generalize.

3.5.4   Analysis of the energy-conversion loop

During a typical motoring stroke the locus of the operating point $[i, \psi]$ follows a curve similar to the one in Figure 3.23, which is drawn together with the aligned and unaligned magnetization curves and another magnetization curve at the commutation angle C. At this point the supply voltage is reversed and the current freewheels through the diode. At C the accumulated energy from the supply is equal to the total area $U = W_{mt} + W_{fC}$. The stored magnetic energy is equal to W_{fC}. Therefore the mechanical work done between turn-on and commutation is W_{mt}, during the period of transistor conduction. Note that in Figure 3.23 this is roughly comparable to W_{fC}, meaning that only half of the energy supplied has been converted to mechanical work. The other half is stored in the magnetic field.

After commutation, Figure 3.24, the supply voltage is reversed and the energy W_d is returned to the supply *via* the diodes. The mechanical work done during the freewheeling interval is $W_{md} = W_{fC} - W_d$. In Figure 3.24 this is less than one-half of W_{fC}. A simple energy balance can be deduced from the estimated areas in Figures 3.23 and 3.24. Suppose that the energy supplied from the controller during the 'fluxing' interval (transistor conduction) is $U = W_{mt} + W_{fC} = 10$ joules. At C,

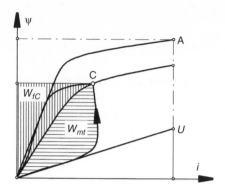

Fig. 3.23 Analysis of energy-conversion loop: transistor conduction.

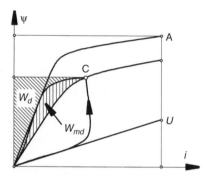

Fig. 3.24 Analysis of energy-conversion loop: diode freewheeling.

5 J have been converted to mechanical work and 5 J are stored in the field. During the 'de-fluxing' period (diode freewheeling), $W_d = 3.5$ J is returned to the supply and $W_{md} = 1.5$ J is converted to mechanical work. The total mechanical work is therefore $W = W_{mt} + W_{md} = 5 + 1.5 = 6.5$ J or 65% of the energy originally supplied by the controller. The energy returned to the supply is $W_d = 3.5$ J or 35% on each 'stroke'.

The entire stroke is shown in Figure 3.25, which combines the two previous diagrams. The energy conversion is now shown as the area W, while the energy

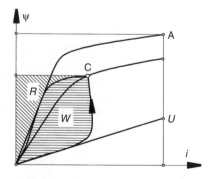

Fig. 3.25 Analysis of energy-conversion loop: the combined loop.

returned to the supply is $R = W_d$. The original energy supplied by the controller is $U = W + R$, and the energy ratio is 0.65.

3.6   Obtaining the magnetization curves

This subject is fairly well covered in the literature and only an outline or guide is included here, since the main focus of this book is the electronic control (including simulation).

3.6.1   Calculation

The aligned magnetization curve can be calculated by a straightforward lumped-parameter magnetic circuit analysis, with an allowance for the stator slot-leakage which is especially significant at high flux levels. The unaligned curve is more difficult to calculate because of the complexity of the magnetic flux paths in this position, but quite practical results are reported by Miller and McGilp (1990) using a dual-energy method based on a quadrilateral discretization of the slotted region. Earlier work by Corda and Stephenson (1979) also produces adequate results for many practical or preliminary design calculations.

The computation should include the 'partial linkage' effect, meaning that at certain rotor positions the turns of each coil do not all link the same flux. In a finite-element calculation this implies that the flux-*linkage* should be calculated as the integral of $A \cdot dl$ along the actual conductors, with the coilsides in the finite-element mesh in their correct positions in the slot.

End effects are important in switched reluctance motors (see Michaelides and Pollock (1994), Reece and Preston (2000)). When the rotor is at or near the unaligned position, the magnetic flux tends to 'bulge out' in the axial direction. The associated increase in the magnetic permeance can raise the unaligned inductance L_u by 20–30%. Since this inductance is critical in the performance calculations, it is important to have a reasonable estimate of it. Unfortunately two-dimensional finite-element analysis cannot help with this problem, and three-dimensional finite-element calculations tend to be expensive and slow. When the rotor is at or near the aligned position, the flux is generally higher and the 'bulging' of flux outside the core depends on the flux level in the laminations near the ends of the stack. At or near the aligned position at high flux levels, the stator and rotor poles can be highly saturated and the external flux-paths at the ends of the machine can increase the overall flux-linkage by a few per cent.

In spite of the complexity of the field problem, good results have been obtained with relatively simple end-effect factors for L_u. For example, the *PC-SRD* computer program (Miller and McGilp, 1999) splits the end-effect calculation into two parts. For L_u,

$$L_u = L_{u0} \times (e_u + f_u) \tag{3.35}$$

where $e_u = L_{end}/L_{u0}$ represents the self-inductance L_{end} of the end-windings (including any extension), expressed as a fraction of the uncorrected two-dimensional unaligned inductance L_{u0}. L_{end} is the inductance of a circular coil whose circumference is equal to the total end-turn length of one pole-coil, including both ends. It is multiplied by the

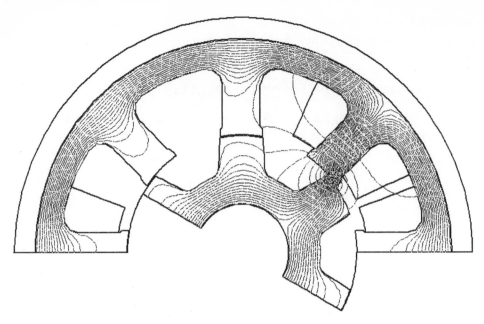

Fig. 3.26 Finite-element flux-plot in a partial-overlap position.

appropriate function of turns/pole and parallel paths before being normalized to L_{u0}. f_u is a factor that accounts for the *axial* fringing in the end-region. It is calculated by the approximation

$$f_u = \frac{d_r + L_{stk}}{L_{stk}} \tag{3.36}$$

which is derived by analogy with the fringing formula for two opposite teeth or poles: d_r is the rotor slot depth and L_{stk} is the stack length. For the aligned position the procedure is similar, with a factor $f_a = (g + L_{stk})/L_{stk}$, g being the airgap length.

At positions intermediate between the aligned and unaligned positions, the calculation of individual magnetization curves is not practical by analytical methods and the finite-element method should be used. Figure 3.1 shows a simple example of a finite-element flux-plot at a position of partial overlap, and Figure 3.26 a more complex example. Considerable success has been achieved with interpolating procedures especially in computer programs for rapid design (Miller and McGilp, 1990; Miller *et al.*, 1998; Miller and McGilp, 1999).

Measurement of the magnetization curves is described in Miller (1993) and Cossar and Miller (1992).

3.7   Solution of the machine equations

3.7.1   *PC-SRD* method

The *PC-SRD* computer program (Miller and McGilp, 1999) has been widely used for design and analysis of switched reluctance machines for many years, and its original procedure is used here as an example of a successful approach to the simulation of the machine equations. It is based on the nonlinear mathematical model described earlier, applied to one phase of the machine. Each phase is treated as a variable inductance in which the flux-linkage ψ is a nonlinear function of both the current i and the rotor position θ: thus $\psi = \psi(i, \theta)$. *PC-SRD* computes the function $\psi(i, \theta)$ from the geometry, the winding details, the *B/H* curve of the steel, and it represents the result graphically as a set of static magnetization curves in which ψ is plotted vs. i at several rotor positions between the unaligned and the aligned positions. For a given machine, the curves are a fixed property. It is not necessary to recalculate them unless changes are made to the geometry, the windings, or the steel. For precise work *PC-SRD* can import external magnetization curves which have been obtained either by measurement or by 3D finite-element calculation.

PC-SRD solves the electrical circuit by integrating (3.1) for one phase by Euler's method on a step-by-step basis. At each integration step there results a new value of flux-linkage ψ, and *PC-SRD* computes a new value of current i from the function $\psi(i, \theta)$, i.e. from the magnetization curves. The method of solving for each new current value i depends on whether the curves are internal or external. With internal magnetization curves *PC-SRD* uses a fast algebraic interpolation method based on so-called *gauge curves* originally described in Miller and McGilp, (1990). With external curves, *PC-SRD* curve-fits the curves with a set of cubic splines, and uses an interpolation process. This is slower than the gauge curve method, but more accurate.

The instantaneous torque is calculated from the rate of change of coenergy $\partial W_c(i, \theta)/\partial\theta$. With internal magnetization curves, the derivative is evaluated using approximate algebraic expressions derived from the gauge curve model. With external mag curves it is evaluated from a precalculated set of spline functions that represent the coenergy W_c as a function of current and rotor position, $W_c(i, \theta)$. The average electromagnetic torque is computed from the loop area W in Figure 3.15, and a typical example is given in Figure 3.16. Several computed examples are given in Chapter 5. The average electromagnetic torque is given by (3.18). Since W is evidently an integral quantity, errors in the distribution of the magnetization curves tend to cancel out, provided that the aligned and unaligned curves are accurate. On the other hand, the calculation of the instantaneous torque is sensitive to the precision in the intermediate magnetization curves and the method of representing them mathematically is critical.

PC-SRD's model is a single-phase model. The currents in phases $2, 3 \ldots$ are determined by phase-shifting the current waveform of phase 1, which is calculated as though it were the only current flowing. The fluxes and torques of the other phases are added to those of phase 1 without taking account of interactions in shared magnetic circuits. This is one of the main limitations of *PC-SRD*, but it also explains the extraordinary speed of computation. A full magnetic model of a polyphase switched reluctance motor, including all magnetic interactions between phases, requires a multi-dimensional set of magnetization curves and is a formidably complex thing to contemplate. The reason why *PC-SRD*'s simple model is so successful in general is that switched reluctance motors are generally designed with sufficient yoke dimensions to avoid mutual interaction between phases. Nevertheless, under conditions of extreme loading or under fault conditions, the *PC-SRD* model cannot be expected to give accurate results, even with external magnetization curves, since these are still valid only for one phase conducting.

3.7.2   A method that uses co-energy and avoids integration

The somewhat convoluted process described in the previous section could in principle be replaced by a more direct method based on a co-energy map. The voltage equation for one phase is

$$v = e + Ri \qquad (3.37)$$

where the back-EMF is given by

$$e = \frac{\partial \psi}{\partial t} = \omega_m \frac{\partial \psi}{\partial \theta} \qquad (3.38)$$

and this can be obtained from the $\psi(i, \theta)$ curves by differentiating with respect to θ at constant current. The solution proceeds at each timestep by calculating

$$i = \frac{v - e}{R} \qquad (3.39)$$

using the current values of v and e. Once the 'new' current is calculated from (3.39), e is re-evaluated using (3.38) for the next integration step. The torque is calculated using

$$T = \frac{\partial W_c}{\partial \theta} \qquad (3.40)$$

which is also evaluated at constant current. It is interesting to note that the flux-linkage can also be evaluated from the co-energy using

$$\psi = \frac{\partial W_c}{\partial i} \qquad (3.41)$$

evaluated at constant θ. This suggests that the back-EMF can be evaluated using

$$e = \omega_m \frac{\partial^2 W_c}{\partial \theta \partial i}. \qquad (3.42)$$

This in turn suggests that the machine can be represented by a surface $W_c(i, \theta)$ whose derivatives can be used at any position θ and any current i to determine the back-EMF e (3.38), the flux-linkage ψ (3.41) and the torque T (3.40). The finite-element solution of the 'magnetization curves' can therefore be expressed in terms of the surface $W_c(i, \theta)$ without even computing the flux-linkage ψ, given that the finite-element method can compute W_c directly by means of a global integration. (This method could probably be applied also to permanent-magnet machines.) A difficulty with this approach, as with the previous one, is the representation of the co-energy function by a sufficiently smooth interpolating function that is differentiable in both θ and i. Also, it does not naturally provide data which can be compared with measurements.

3.8   Summary

In this chapter we have described the basic electromagnetic characteristics of the switched reluctance machine, starting with a simple magnetically linear model and going on to the analysis of the magnetically nonlinear machine. The treatment has been for excitation by one phase, and the total torque has been assumed to be the sum of the phase torques: that is, magnetic coupling between phases has been ignored. In machines with exceptionally high magnetic loading this assumption can lead to a serious overestimate of the torque capability, but the analysis of such cases requires extensive finite-element analysis and is beyond the scope of the book.

The electromagnetic analysis has been developed as a basis for the development of the control strategies described in later chapters, and therefore some attention has been given to the firing angles and the nature of the torque/speed characteristic. The importance of magnetization curves has been emphasized, and an outline has been given of methods of obtaining good magnetization curves as well as the methods of simulation which are used in computer-aided analysis and design of the switched reluctance machine and its controller.

4

Designing for low noise

W.A. Pengov and R.L. Weinberg
MaVRik Motors, Cleveland

This chapter is in two parts. The first three sections introduce the basic principles of noise generation in the switched reluctance machine, and outline some of the main methods of minimizing the noise level.[1] The second part details a successful example of a switched reluctance machine designed for inherently low noise and low torque ripple, the *stagger-tooth*™ motor (Pengov, 1998).

4.1 Noise in the switched reluctance machine

The development of switched reluctance machines since the mid-1960s has coincided with a steady reduction of acoustic noise and torque ripple in almost all variable-speed motor drives. Switched reluctance machines are unfortunately held to have higher noise and torque ripple than brushless permanent-magnet motors or induction motors, both of which have natural electromechanical advantages in these areas. However, the technical and economic advantages of switched reluctance motors continue to make them attractive for a range of applications such as washing-machine drives, high-speed blowers or compressors, general-purpose industrial drives up to a few tens of kW, aerospace actuators, and others. The low-cost end of the market is also one of the most sensitive to acoustic noise because many of the applications are in residential or automotive environments, and therefore it is desirable to find cost-effective ways to exploit the switched reluctance motor without problems arising from noise or torque ripple.

Some perspectives
Although noise minimization is important in many applications, there are still many cases where it is not. For example, in aircraft actuators and certain types of industrial machinery the background noise of the driven mechanism, and/or the ambient noise level, may be so high that a few additional decibels from the drive motor will make little difference.

The switched reluctance machine is not the only machine capable of noisy operation. Induction motors can also be noisy, for example, if the combination of stator

---

[1] These sections were added by the editor.

and rotor slot-numbers is inappropriate. It is easy to forget that it took many years of painstaking experimental and theoretical research to determine design rules for quiet operation of induction machines. Moreover, the induction machine can encounter problems of cogging, crawling, and synchronous locking torques which, when they occur, are arguably more severe than anything liable to arise in the switched reluctance machine.

Because of the flexibility of configuration and design of the switched reluctance machine, it is hardly possible to claim that the noise problem has a general solution that can be expressed as a set of design guidelines. On the contrary, it is more likely at the present state of the art, that each design will require individual attention to its acoustic properties. Allowance should therefore be made for some flexibility in the design, and particularly in the control of the current waveform, to alleviate noise problems if and when they arise.

4.2   Noise-generation mechanisms and characterization

The most important source of noise in the switched reluctance machine is the distortion of the stator core by predominantly radial magnetic forces, Figure 4.1(a) (Cameron *et al.*, 1992; Colby *et al.*, 1996; Besbes *et al.*, 1998). If the number of working pole-pairs is two, as in a four-phase 8/6 machine, the distortion is an ovalization. If the number of working pole-pairs is four, it is as shown in Figure 4.1(b). The stiffness of the core is significantly greater for the 4-pole case than the 2-pole case. For the 2-pole case it is proportional to the ratio $(R/t)^3$, where t is the radial thickness of the stator yoke and R is its mean radius. The ideal distribution of radial forces would be uniform, as in Figure 4.1(c). It must be observed that the forces are pulsed, so that if a distortion is excited at its natural resonant frequency, the displacement required to produce a significantly loud noise may require only a small current. It follows, however, that only a small modulation of the current may be necessary to cancel this resonance (see Chapter 8).

The distortions in Figure 4.1 are only the fundamental ones; that is, those with the lowest resonant frequency. A more detailed analysis of the eigenmodes of vibration reveals several others which are capable of contributing significantly to the noise level, and some of these are illustrated for a four-phase 8/6 motor in Figure 4.2 (Colby *et al.*, 1996). This suggests that the cancellation of noise by means of harmonic injection in the current waveform may require attention to more than one frequency.

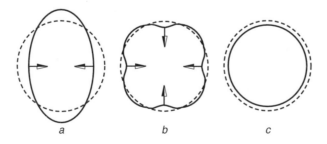

Fig. 4.1 Basic vibration modes.

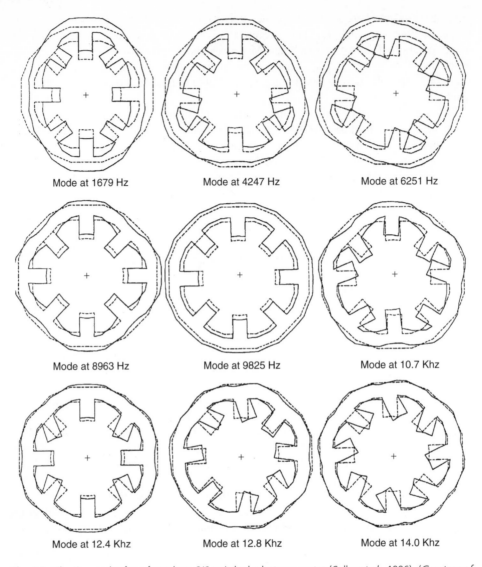

Fig. 4.2 Vibration modes for a four-phase 8/6 switched reluctance motor (Colby *et al.*, 1996). (*Courtesy of IEEE*).

4.2.1 Noise characterization

Acoustic noise is characterized by factors designed to relate it to human perception, taking into account the nonlinear response of the ear. The sound *power* level L_w emitted by a noise source is defined by

$$L_w = 10\log_{10} \frac{W}{W_0} \quad [\text{dB}] \tag{4.1}$$

and is measured in decibels. The base power level is $W_0 = 1\,\mathrm{pW}$. The perceived sound *pressure* level L_p is defined by

$$L_p = 20 \log_{10} \frac{P}{P_0} \quad [\mathrm{dB}]. \tag{4.2}$$

The base pressure level is $P_0 = 20\,\mu\mathrm{P}$. The emitted power level and perceived pressure level are related by (4.3): if R is the distance from the noise source over an area S the pressure is averaged as

$$L_p = L_w - 10 \log_{10} \frac{2\pi R^2}{S} \quad [\mathrm{dB}]. \tag{4.3}$$

For typical values, normal conversation has a power level of about 55 dB; a loud truck might generate over 90 dB, and an aircraft engine over 120 dB.

When there are multiple sources, the power levels are added: thus the effect of two sources M and L is

$$\begin{aligned} M + L &= 10 \log_{10}[10^{M/10} + 10^{L/10}] \\ &= M + 10 \log_{10}[1 + 10^{(L-M)/10}]. \end{aligned} \tag{4.4}$$

For example, if $M = 87\,\mathrm{dB}$ and $L = 90\,\mathrm{dB}$, $M + L = 91.76\,\mathrm{dB}$. This illustrates the fact that two noisy sources of roughly the same sound power level do not produce the perception of 'twice the noise'.

A further refinement in noise measurement is to filter the measured noise in such a way as to take into account the frequency-variation of the sensitivity of the human ear. When this is done, the noise is measured in 'dBA'. Figure 4.3 shows the standard A-weighting curve. Permitted noise levels are of course regulated by standards, an example of which is shown in Table 4.1 from the IEC.

4.2.2   Testing

In practice the on-load noise is of most interest, but the load contributes noise and therefore testing and specifications are usually at no-load, sometimes with a no-load correction (e.g. 4 dBA for 4-pole induction motor). A switched reluctance machine

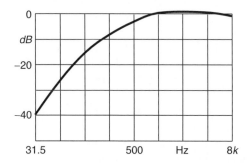

Fig. 4.3 A-weighting curve.

Table 4.1 Example of IEC noise regulations

Motor rated power, kW	Maximum permissible A-weighted sound power levels in dBA from IEC 34-6 (either or both air inlets are open: i.e. enclosures IC01, IC11, IC21)					
	Speed, rpm					
	<960	960–1320	1320–1900	1900–2360	2360–3150	3150–3750
1–1.1	73	76	77	79	81	82
1.1–2.2	74	78	81	83	86	86
2.2–5.5	77	81	85	86	89	93
5.5–11	81	85	88	90	93	97
11–22	84	88	91	93	96	97

may have a much bigger difference in noise level between no-load and on-load, than induction or other motors; and if this is so, the standards would appear to be favorable to the switched reluctance machine (and unfavorable to the user!). Tests usually are conducted in 'free-field' conditions, but if detailed analysis is required an anechoic chamber may be used. The sound power level is measured at several positions around the motor, background noise is subtracted out, then A-weighting is applied.

4.3 Methods for avoiding noisy operation

4.3.1 The 39 steps

The following '39 steps' have been compiled as a basic list of advisable measures to avoid noisy operation, but of course many more could be added. It is difficult to assign priorities in general.

1. Avoid the temptation to design for extremely high power density, unless it is absolutely necessary.
2. Design the machine with a higher 'electric loading' and a lower 'magnetic loading'.
3. Use a pole combination with more than one pair of working poles, i.e. with a 'multiplicity' greater than 1. For example, a 12/8 machine ought to be less noisy than an 8/6 machine.
4. Make sure the lamination steel is truly anisotropic and has no trace of grain orientation. It may help in this regard to rotate successive laminations by one stator pole-pitch as they are stacked. A relatively low-permeability lamination steel may also help to reduce noise level.
5. Use the thickest possible stator yoke. This maximizes the stiffness against deflections of the type shown in Figure 4.1, and against many others shown in Figure 4.2.
6. Use the largest airgap size that can be tolerated without undue loss of inductance ratio.
7. Make the rotor and stator poles as wide as possible, with the rotor poles slightly wider than the stator poles.
8. Use a slight taper on the rotor and stator poles.
9. Use a fillet radius at pole-root corners on the stator and the rotor (where the pole joins the yoke).

10. The rotor pole leading edges can be tapered or shaped very slightly to reduce the harmonic content in the instantaneous torque waveform.
11. Avoid notches and holes in the laminations.
12. Laminations should be annealed after stamping.
13. It may help to use pre-coated laminations.
14. Make sure the laminations are truly symmetric. It is possible with some production methods to get angular errors in the pole locations: for example, in a 6-pole stator two poles might not be exactly opposite.
15. Use the largest possible shaft diameter. The rotor slot-depth may need to be made shallower to permit a larger shaft to be accommodated without diminishing the rotor yoke thickness. When the rotor slot-depth is reduced, the unaligned inductance increases, but it may be worth losing a little on the inductance ratio to get a stiff shaft.
16. If possible, encapsulate the rotor to reduce windage noise. Obviously this must be done in such a way as to support the potting material against centrifugal loads at high speed.
17. Locate the bearings as close as possible to the rotor stack. The end-brackets carrying the bearings may sometimes protrude under the end-windings to get the bearings close in.
18. Use large-ball antifriction bearings, with the largest feasible diameter and a tight tolerance. Sleeve bearings are thought to be unsuitable although opinions differ on this point.
19. Make sure the rotor is axially centred (magnetically, not necessarily mechanically). There should be enough end-float to allow centring, but not too much, and one bearing should be spring-loaded (use a wavy washer).
20. Use an unfinned frame, preferably aluminium (thin rolled steel may resonate and adds little radial stiffness).
21. Make sure the end-brackets are rigidly attached to the frame. A tapered flange or spigot helps this.
22. Ensure electrical and magnetic balance between the coils of each phase (resistance and inductance). This includes leads and internal connections between coils.
23. If possible, use parallel connection of pole-coils, not series. This compensates for airgap inconsistencies by allowing greater current through the poles with the larger gap.
24. Use the highest number of turns consistent with achieving the required torque/speed curve at the available voltage. This reduces the speed at which the control can switch over to single-pulse operation with phase angle control; this is quieter than low-speed chopping.
25. Clamp up the lamination stacks if possible, under pressure. An alternative is to impregnate them with varnish (after winding), preferably by vacuum impregnation.
26. Use nonmetallic slot-wedges (top-sticks) to close the stator slots after winding, and double-thickness slot liners. This adds rigidity and damping to the stator.
27. Tie the end-windings with cord before varnishing.
28. Maintain the best possible concentricity between the rotor and stator. This requires precise assembly fixturing and may warrant grinding on centres that are carefully maintained concentric.
29. Wind the coils as tightly as possible.

30. If possible, encapsulate the stator windings.
31. A very small amount of skew may be used on the rotor (or stator).
32. Voltage-PWM in the drive is known to produce lower noise than hysteresis-band current regulation.
33. With voltage-PWM it may also help to dither the duty-cycle and/or frequency.
34. Even lower noise can be achieved with a variable-voltage d.c. source, such that the phase transistors are used only for commutation and controlling the firing angles.
35. The current should be chopped with only one transistor, not both ('soft chopping').
36. Adjust the firing angles on test to give the quietest operation at a given load. Even small changes make a marked difference. There may be a trade-off with efficiency.
37. Current-profiling may help, but requires extensive analysis and experimentation, and may be complicated.
38. Use the highest possible chopping frequency to minimize ripple current and, if possible, set the chopping frequency above the human audible range (typically $>15–20\,\mathrm{kHz}$).
39. If possible, mount the machine on flexible antivibration mountings, and use a flexible rubber-bush type coupling to the load.

4.3.2 Commutation control

It was recognized by Horst (1995) and also by Pollock (1997) that a significant component of acoustic noise is attributable to the excitation of a stator vibration mode at the point of commutation (switching off) of the phase current at the end of each transistor conduction period. Commutation generally occurs when the active stator and rotor poles are substantially aligned, and at this position the current and flux-density are both high, resulting in large radial forces. Given that an eigenmode can be excited not only by the force, but also by its time-derivatives, the shape of the current waveform at commutation assumes a critical significance. Accordingly a reduction in the noise can be effected by 'softening' the current profile at the point of commutation, Figure 4.4.

The method shown in Figure 4.4 is to commutate in two stages, by switching off one of the two phaseleg transistors before the other, producing a zero-volt loop between c1 and c2. The energy-conversion loop in Figure 4.4 shows that this need not incur any significant loss of torque.

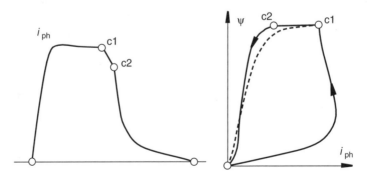

Fig. 4.4 Two-stage commutation.

Another method which is also applied at the point of commutation is to phase the switching of the transistors so precisely as to cancel the main vibration mode, as described in Pollock (1997) and in more detail in Chapter 8.

4.4 The stagger-tooth™ motor

The stagger-tooth motor (Pengov, 1998; Pengov and Weinberg, 1999) has a two-stage flux pattern which arises from its geometry, producing a natural dispersal of the ovalizing stress which leads to a marked noise reduction without the need for current-profiling. The stagger-tooth™ motor was devised to produce low torque ripple in a low-cost, two-phase motor intended mainly for unidirectional applications. Early switched reluctance motors used asymmetrical pole geometry on the rotor (e.g. Byrne and Lacy,(1976))• to reduce torque 'dips' or 'dead spots' rather than to reduce acoustic noise. In virtually all cases the rotor poles are symmetrically disposed around the rotor, but in the stagger-tooth™ motor adjacent rotor poles have unequal widths, resulting in the two-stage flux-pattern.

The principle is described using the linearized model in which the electromagnetic torque $T = \frac{1}{2}i^2 \, dL/d\theta$ is proportional to the rate of change of phase self-inductance L with respect to rotor position θ, Figure 4.6.

The basic stagger-tooth™ motor has a regular 8-pole stator with four poles/phase and all coils in series. Alternate poles are wound with different phases to produce an excitation pattern N0N0S0S0 from phase A and 0N0N0S0S or 0S0S0N0N from phase B. Two opposite rotor poles have a pole-arc β_r, with $\beta_r \simeq \beta_s$, while the alternate pair have a pole-arc $\sigma_r > \beta_s$. The inductance profile can be constructed by considering the two pairs of stator poles separately. If the rotor begins to rotate CCW from the position in Figure 4.5, poles A1–A3 will begin to overlap with the wide rotor poles and their flux-linkage will contribute to the total phase inductance according to the top trace in Figure 4.6. Torque is initially produced by the overlap between poles A1–A3 and the wide rotor poles, and this continues for a rotation angle of approximately β_s. At

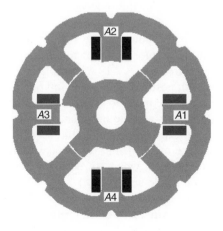

Fig. 4.5 General outline of stagger-tooth™ motor (Pengov, 1998).

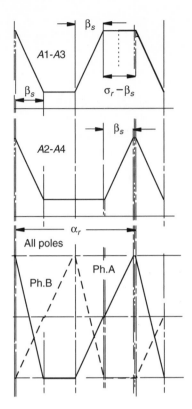

Fig. 4.6 Linearized inductance profiles of stagger-tooth™ motor.

this point (assuming constant phase current) the flux-linkage of poles A1–A3 becomes constant and the torque contributed by these poles falls to zero. If the pole-arcs are correctly chosen, however, the overlap between poles A2–A4 and the narrow rotor poles begins to increase at this instant, and the torque production is taken over by poles A2–A4, whose contribution to the inductance profile is shown in the middle trace in Figure 4.6.

The total inductance profile is shown in the bottom trace in Figure 4.6, and an example measured on a particular motor is shown in Figure 4.7. If the pole-arcs are chosen correctly the transition from poles A1–A3 to poles A2–A4 is smooth, resulting in a wide range of rotation with positive $dL/d\theta$. The full range of nonnegative $dL/d\theta$ is $\sigma_r - (\beta_r - \beta_s)$, with a 'ledge' at the transition point of width $\sigma_r - (\beta_s + \beta_r)$. If $\beta_r = \beta_s$ and $\sigma_r = 2\beta_s$, the ledge disappears and the inductance profile is positive and constant throughout the range $\sigma_r = 2\beta_s$, which is double the normal range in a symmetrical or 'regular' switched reluctance motor. (In a 'regular' switched reluctance motor $L(\theta)$ is symmetrical, and the range of θ in which $dL/d\theta > 0$ is approximately equal to the narrower of the stator and rotor pole-arcs β_s, β_r.)

In the 8/4 stagger-tooth™ motor the rotor pole-pitch is $360/4 = 90°$ and each phase must be capable of producing maximum torque/ampere over at least half of this range, i.e. 45°. This requires that the pole-arcs β_s and β_r be approximately 22.5° while $\sigma_r = 45°$. The gaps between the corners of two consecutive rotor poles therefore have the two

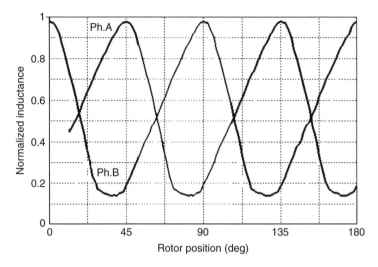

Fig. 4.7 Measured inductance profiles of stagger-tooth™ motor.

values $90 - (45 + 22.5)/2 = 56.25°$ and $90 - 2 \times 22.5/2 = 67.5°$. In the minimum-inductance position the clearance between the corners of an excited stator pole and the nearest rotor pole is therefore at least $(56.25 - 22.5)/2 \simeq 17°$, which ensures a low 'unaligned' inductance and a high inductance ratio. The 'dwell' at the minimum inductance is $\alpha_r - (\sigma_r + \beta_s) = 22.5°$, while the range of positive $dL/d\theta$ is $45°$.

Figure 4.8 shows the flux patterns at three positions, illustrating the transition from 2-pole to 4-pole patterns and short flux-paths.

Actual motor test results of this configuration reveal the performance, using static torque measurements, at various angles through a half revolution. Figure 4.9 shows an example of these results. A constant current was driven through phase A while the rotor was slowly rotated. Torque was measured and plotted at each angular position. Simulated phase B torque curves are superimposed. With 50% dwell for each phase, the natural phase overlap due to the current decay time fills in the torque dips between phases quite well. A small dip in torque can be seen mid-stroke in the torque waveform due to saturation in the first pair of stator teeth. At higher current levels this dip becomes more prominent and contributes to torque ripple.

Fig. 4.8 Series of finite-element flux-plots of stagger-tooth™ motor during one stroke.

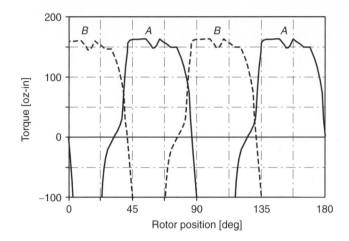

Fig. 4.9 Static torque curves of stagger-tooth™ motor, showing overlap between phases.

Fig. 4.10 Components of a two-phase 16/8 stagger-tooth™ motor. (Courtesy of MaVRik Motors, Inc.).

At low speeds the stagger-tooth™ motor has additional copper loss compared with the equivalent four-phase motor because only two of the four excited poles are producing torque during the early part of the stroke. However, this trade-off is justified in some applications by the improvement in noise level and the reduction in the drive electronics, which has only half the number of switching devices and connectors.

Figure 4.10 shows components of a 16/8 stagger-tooth™ motor which operates on the same principle as the 8/4, but since it has a 'multiplicity' of 2 the flux pattern changes from 8-pole to 4-pole during the stroke.

Stagger-tooth™ designs have been developed for several applications including floor-care appliances, fans, blowers and pumps, and even traction. In an example of a floor-buffing machine, its higher efficiency resulted in more power to the floor with the same supply current limitation of 15 A. An electric motorcycle traction motor ran at speeds up to 14 000 rev/min producing a peak power of 8 HP. Both of these applications would be intolerant of the torque dips exhibited by the typical two-phase switched reluctance motor.

4.5 Summary

This chapter has reviewed some of the main sources of noise in switched reluctance machines, particularly the ovalization and other eigenmodes of vibration in the stator core. The definitions of standard noise measurement parameters used in connection with electric machines have been summarized, and it has been pointed out that standards based on no-load measurement may be favorable to the switched reluctance machine (but unfavorable to the user). A list of '39 steps' which may help with noise reduction has been presented.

The stagger-tooth™ motor has been described in some detail. This machine has not only a low noise level but also a remarkably smooth torque waveform for a two-phase switched reluctance motor, and uses a smaller number of drive components than equivalent three-phase or four-phase machines. These advantages justify the small loss of efficiency arising from the fact that not all poles are active for the entire stroke.

Finally it is noted that in many applications, acoustic noise is not of primary concern – for example, in aircraft actuators and certain types of industrial machinery, the background noise level is so high as to render the additional noise of the motor inconsequential.

5

Average torque control

Lynne Kelly; Calum Cossar and T.J.E. Miller
Motorola; SPEED Laboratory, University of Glasgow

5.1 Introduction

Torque in the switched reluctance machine is produced by pulses of phase current synchronized with rotor position. The timing and regulation of these current pulses are controlled by the drive circuit and the torque control scheme. Usually there are also outer feedback loops for controlling speed or shaft position, as shown in Figure 5.1. The outer loops are generally similar to those used in other types of motor drive, but the inner torque loop is specific to the switched reluctance machine.

The torque demand signal generated by the outer control loops is translated into individual current reference signals for each phase (Bose, 1987). The torque is controlled by regulating these currents. Usually there is no torque sensor and therefore the torque control loop is not a closed loop. Consequently, if smooth torque is required, any variation in the torque/current or torque/position relationships must be compensated in the feedforward torque control algorithm. This implies that the torque control algorithm must incorporate some kind of 'motor model'. Unlike the d.c. or brushless d.c. motor drive, the switched reluctance motor drive cannot be characterized by a simple torque constant k_T (torque/ampere). This in turn implies that the drive controller must be specifically programmed for a particular motor, and possibly also for a particular application. It also implies that one cannot take a switched reluctance motor from one source and connect it to a drive from another source, even when the voltage and current ratings are matched. On the contrary, the motor and drive control must be designed together, and usually they must be optimized or tuned for a particular application.

The power electronic drive circuit is usually built from phaselegs of the form shown in Figure 3.6. These circuits can supply current in only one direction, but they can supply positive, negative, or zero voltage at the phase terminals. Each phase in the machine may be connected to a phaseleg of this type, and the phases together with their phaseleg drive circuits are essentially independent. The circuits in Figure 3.7 make it possible to operate the phases with separate d.c. supplies of different voltages, although the most usual case is to connect them all to a common d.c. supply. Figures 3.7(a) and 3.7(b) also show the possibility of 'fluxing' at one voltage V_1 and 'de-fluxing' at another voltage $-V_2$.

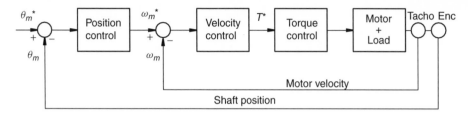

Fig. 5.1 Nested control loops. T^* = torque demand; ω_m^* = speed demand, ω_m = speed; θ^* = position demand; θ = shaft position. Tacho = tachometer or speed transducer; Enc = encoder or position transducer.

At lower speeds the torque is limited only by the current, which is regulated either by voltage-PWM ('pulse-width modulation'), or instantaneous current regulation. As the speed increases the back-EMF increases to a level at which there is insufficient voltage available to regulate the current; the torque can then be controlled only by the *timing* of the current pulses. This control mode is called 'single-pulse mode' or 'firing angle control', since the firing angles alone are controlled to produce the desired torque. Many applications require a combination of the high-speed and low-speed control modes. Even at lower speeds with voltage-PWM or current regulation, the firing angles are typically scheduled with speed to optimise performance (Lawrenson, 1980).

This chapter is concerned with control of *average* torque.[1] The simplest definition of 'average torque' is the torque averaged over one stroke ($\varepsilon = 2\pi/mN_r$). The amplitude and phase of the current reference signal (relative to the rotor position) are assumed to remain constant during each stroke. Loosely speaking, this corresponds to the operation of a 'variable-speed drive', as distinct from a servo drive which would be expected to control the instantaneous torque (see Chapter 6). Average torque control requires a lower control-loop bandwidth than instantaneous torque control.

Differences between switched reluctance machines and classical machines

Much of the classical theory of torque control in electric drives is based on the d.c. machine, in which torque is proportional to flux × current. The flux and current are controlled independently, and the 'orientation' of the flux and the ampere-conductor distribution, both in space and in time, is fixed by the commutator. In a.c. field-oriented control, mathematical transformations are used, in effect, to achieve independent control of flux and current, and the commutator is replaced by a shaft-position sensor which is used by the control processor to adjust the magnitude and phase of the currents to the correct relationship with respect to the flux. The current can be varied very rapidly so that a rapid torque response can be achieved. Generally speaking, in classical d.c. and a.c. machines the flux is maintained constant while the current is varied in response to the torque demand. In both cases the torque control theory is characterized by the very important concept of 'orthogonality', which loosely means that the flux and current are 'at right angles'. In the architecture of the machine and the drive, this concept has a precise mathematical meaning which depends on the particular form or model of the system.

---

[1] Also known as 'running torque'; as distinct from instantaneous torque.

In switched reluctance machines unfortunately, there is no equivalent of field-oriented control. Torque is produced in impulses and the flux in each phase must usually be built up from zero and returned to zero each stroke. The 'orthogonality' of the flux and current is very difficult to contemplate, because the machine is 'singly excited' and therefore the 'armature current' and 'field current' are indistinguishable from the actual phase current. Although this appears to be the case also with induction machines, the induction machine has sine-distributed windings and a smooth airgap, so that the theory of space vectors can be used to resolve the instantaneous phase currents into an MMF distribution which has both direction and magnitude, and the components of this MMF distribution can be aligned with the flux or orthogonal to it. The switched reluctance machine does not have sine-distributed windings or a smooth airgap, and there is virtually no hope of 'field-oriented' control. To achieve continuous control of the instantaneous torque, the current waveform must be modulated according to a complex mathematical model of the machine: see Chapters 6, 8 and 9.

5.2   Variation of current waveform with torque and speed

The average electromagnetic torque is given by (3.18), and the energy-conversion loop area W is shown in Figures 3.15 and 3.16. The objective of 'average torque control' is to achieve a reasonably simple current pulse waveform which produces the required value of W corresponding to the torque demand. Even in simple cases, this is more complex than simply determining the required 'value of current', since the torque/ampere varies as a function of both position and current. The following sections describe the general properties of the current waveform at different points in the torque/speed diagram, Figure 5.8. This will provide a physical basis for understanding the design and operation of the controller.

5.2.1   Low-speed motoring

At low speed the motor EMF e (3.2, 3.38) is low compared to the available supply voltage V_s, and the current can be regulated by chopping. If voltage-drops in the semiconductor devices are neglected, the drive can apply three voltage levels $+V_s$, $-V_s$ or 0 to the winding terminals, and the voltage difference $(V_s - e)$, $(-V_s - e)$, or 0 is available to raise or lower the flux and current, since $v - e = d\psi/dt$. A simple strategy is to supply constant current throughout the torque zone, that is, over the angle of rotation through which the phase inductance is substantially rising. Figure 5.2 shows a typical low-speed motoring current waveform of this type in a three-phase 6/4 motor at 500 rev/min.

The current waveform i is chopped at about 8 A, starting 5° after the unaligned position (which is at 45°) and finishing 10° before the aligned position (which is at 90°). At first hardly any torque is produced because the inductance is low and unchanging, but when the approaching corners of the stator and rotor poles are within a few degrees of conjunction J, the torque suddenly appears. Clearly the instantaneous torque is controlled by regulating the current. When the driving transistors are switched

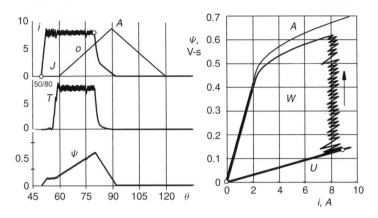

Fig. 5.2 Low-speed motoring waveforms. i = phase current, ψ = phase flux-linkage, T = phase torque, and o = overlap between stator and rotor poles. Horizontal axis is rotor angle (degrees). Unaligned position $U = 45°$; aligned position $A = 90°$. The 'conjunction' position J is the start of overlap between the active rotor poles and the stator poles of this phase.

off, $10°$ before the aligned position, the current commutates into the diodes and falls to zero, reaching the 'extinction' point a few degrees beyond this position, so that virtually no negative torque is produced.

The flux-linkage ψ grows from zero and falls back to zero every stroke. When the driving transistors are first switched on, ψ grows linearly at first because the full supply voltage is applied across the winding terminals. When the current regulator starts to operate, ψ is also regulated to a constant value at first because the constant current is being forced into an inductance that is still almost constant at the low value around the unaligned position, before the poles begin to overlap. As soon as the pole corners approach conjunction J, the inductance starts to increase, so the flux-linkage ψ also increases as constant current is now being forced into a rising inductance. The flux-linkage continues to increase until the commutation point. After that, the diodes connect a negative 'de-fluxing' voltage $-V_s$ across the winding terminals and therefore ψ falls to zero very rapidly. In this example the resistive voltage-drop is small, and therefore the rate of fall of flux-linkage is almost linear. At low speed the dwell is made approximately equal to β_s, since this is 'width' of the 'torque zone', and this angle might typically be a little less than $30°$ in a typical 6/4 motor. De-fluxing is completed over only a small angle of rotation since the speed is low, so the entire conduction stroke occupies only about $30°$.

The process is summarized in the energy-conversion loop, where the energy conversion area W is clearly bounded by the dynamic loop. It is traversed in the counterclockwise direction. The loop fits neatly between the aligned and unaligned magnetization curves as a result of the selection of the firing angles. It appears that the area W could be increased slightly, by retarding the commutation angle to extend the loop up to the aligned magnetization curve. This would not require any increase in peak current, but it would increase the average and r.m.s. values of the current. It is also possible that delayed commutation could incur a period of negative torque just after the aligned position, which would appear as a re-entrant distortion of the energy-conversion loop, so the apparent gain in torque might not be so much as it appears.

Operation is at point M1 in the torque/speed characteristic, Figure 5.8. It is possible to maintain torque constant with essentially the same current waveform as the speed increases up to a much higher value, since the motor EMF is still much lower than the supply voltage.

5.2.2   High-speed motoring

At high speed the motor EMF is increased and the available voltage may be insufficient for chopping, so that the torque can be controlled only by varying the firing angles of a single pulse of current. Figure 5.3 shows a typical example, in which the speed is 1300 rev/min.

The driving transistors are switched on at 50° and off at 80°, the same as in Figure 5.2. At first the overlap between poles is small, and the supply voltage forces an almost linear rise of current $di/dt = V_s/L_u$ into the winding. Just before the start of overlap the inductance begins to increase and the back-EMF suddenly appears, with a value that quickly exceeds the supply voltage and forces di/dt to become negative, making the current fall. The higher the speed, the faster the current falls in this region. Moreover, for a given motor there is nothing that can be done to increase it, other than increasing the supply voltage. The torque also falls. Operation is at point M2 in Figure 5.8.

5.2.3   Operation at much higher speed

At a certain 'base speed' the back-EMF rises to a level at which the transistors must be kept on throughout the stroke in order to sustain the rated current. Any chopping would reduce the average applied voltage and this would reduce the current and torque. The 'base' speed is marked B in Figure 5.8. If resistance is ignored, the peak flux-linkage achieved during the stroke is given by $V_s\Delta\theta/\omega$, where $\Delta\theta$ is the 'dwell' or conduction angle of the transistors. If the peak flux is to be maintained at higher speeds, the

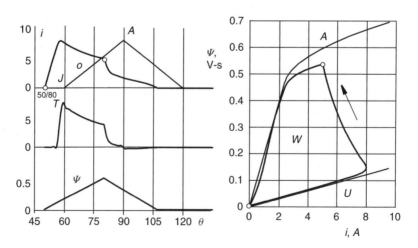

Fig. 5.3 High-speed motoring waveforms.

'dwell' must be increased linearly with speed above the base speed. At high speed the turn-on angle can be advanced at least to the point where the sum of the fluxing and de-fluxing intervals is equal to the rotor pole-pitch, at which point conduction becomes continuous (i.e. the current never falls to zero). This corresponds to a dwell of 45° and a total conduction stroke of 90°, neglecting the effect of resistance (which tends to shorten the de-fluxing interval).

Thus it appears that the dwell or 'flux-building angle' can increase from 30° at low speed to 45° at high speed, an increase of 50% or 1.5:1. Over a speed range of 3:1, the peak flux-linkage might therefore fall to $1.5/3 = 0.5$, or one-half its low-speed value. This is illustrated in Figure 5.4 for a speed of 3900 rev/min. The peak current is approximately unchanged but the loop area W is only about one-third of its low-speed value. The comparison between the loop areas at 1300 and 3900 rev/min is shown more clearly in Figure 5.5. The average torque is therefore only about one-third of its low-speed value, but the *power* remains almost unchanged. Operation is at point M3 in Figure 5.8.

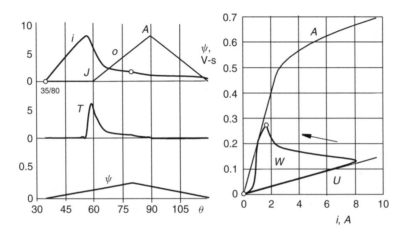

Fig. 5.4 Very high-speed motoring.

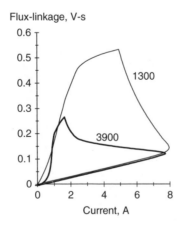

Fig. 5.5 Energy-conversion loops at low and high speed, 1300 and 3900 rev/min.

5.2.4   Low-speed generating

Low-speed generating is similar to low-speed motoring except that the firing angles are retarded so that the current pulse coincides with a period of falling inductance. Figure 5.6 shows a typical example.

The average torque is negative and the energy-conversion loop is traversed in the clockwise direction. At the start of the stroke, there is a slight positive torque because the current is switched on a few degrees before the aligned position, while the inductance is still rising. In this example the torque falls to zero before the current is commutated, indicating that the commutation angle could be advanced a few degrees without reducing the average torque or the energy conversion. The efficiency would improve because the copper loss would be reduced. During that 'tail' period when there is current but no torque, the current is maintained by the drive which is simply exchanging reactive energy with the d.c. link filter capacitor. Operation is at point G1 in Figure 5.8.

5.2.5   High-speed generating

High-speed generating is similar to high-speed motoring, except that again the firing angles are retarded so that the current pulse coincides with a period of falling inductance. Figure 5.7 shows a typical example.

The torque is negative and the energy-conversion loop is again traversed in the clockwise direction. At the start of the stroke, there is a slight positive torque because the current is switched on a few degrees before the aligned position, while the inductance is still rising. Operation is at point G2 in Figure 5.8.

5.2.6   Operating regions – torque/speed characteristic

For control purposes the torque/speed envelope can be divided into regions as shown in Figure 5.8.

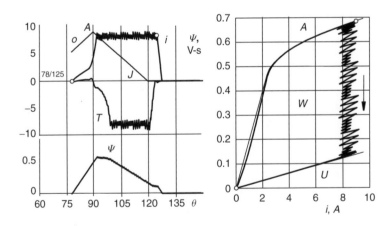

Fig. 5.6 Low-speed generating waveforms.

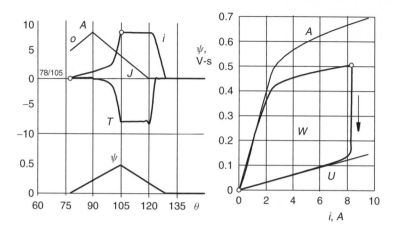

Fig. 5.7 High-speed generating waveforms.

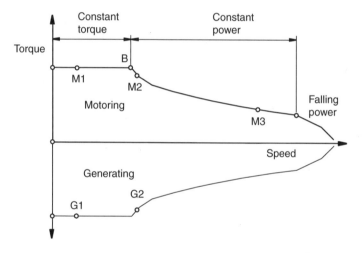

Fig. 5.8 Torque/speed characteristics.

Constant torque region

The base speed is the maximum speed at which maximum current and rated torque can be achieved at rated voltage. In this region the torque is controlled by regulating the current, with relatively minor adjustments in the firing angles as necessary to alleviate noise or improve the current or torque waveform, or to improve efficiency.

Constant power region

As the speed and back-EMF increase, the dwell is increased to maintain the peak flux-linkage at the highest possible level. If the dwell is equal to half the rotor pole-pitch and the de-fluxing angle is negligible at the base speed, then in principle the dwell can be doubled before the onset of continuous conduction. Therefore if the dwell is increased in proportion to speed, the peak flux-linkage can be maintained up to about twice the base speed. However, constant power can be maintained to a higher speed than this, because the loss of loop area $dW/d\omega$ is compensated by the increase in speed. If power is taken as ωT and $T \propto W$, then $P \propto \omega W$ and for constant power we require that $\Delta P = \omega \Delta W + W \Delta \omega = 0$, which says that constant power can be maintained up to the point where $\Delta W/W = -\omega/\Delta \omega$. In other words, the maximum speed at constant power is the speed at which the rate of loss of loop area is balanced by the rate of increase of speed.

It might be of interest to observe that the rate of increase in back-EMF is less than proportional to the speed, because the current decreases with speed and $\partial\psi/\partial\theta$ is therefore also reduced. (In the linear analysis $e = i\omega \, dL/d\theta$, and i is decreasing while ω is increasing and $dL/d\theta$ remains constant.)

Falling power region

Eventually as the speed increases, the turn-on angle can be advanced no more, and the torque falls off more rapidly so that constant power cannot be maintained, even though very high speeds can be attained against a light load. The maximum phase advance depends on the drive controller. If the turn-on angle is advanced beyond the point where the dwell becomes equal to about half the rotor pole-pitch, continuous conduction will begin: the phase current never falls to zero and the energy-conversion loop 'floats' away from the origin. As it does so, it moves to a region where the separation between the aligned and unaligned curves is increased, and the torque per ampere actually increases. For this reason, operation with continuous conduction is a possible means of increasing the power density, not only at high speeds but even at low speeds. The penalty is an increase in copper loss which is acceptable if there is a greater gain in converted power and the machine can withstand the temperature rise. As a means of achieving the same effect, a d.c. bias winding has been suggested for three-phase motors (Horst, 1995; Miller, 1999; Li, 1998).

Reversibility

Figure 5.8 shows only two quadrants of the torque/speed characteristic, corresponding to motoring and generating (or braking). The direction of rotation is the same in both quadrants. Operation in the opposite direction is symmetrical, provided that the rotor position transducer can provide the correct reference position and direction sense. Then the firing angles for motoring in one direction become generating angles in the reverse direction, at least at low speed. The switched reluctance machine and its drive are thus reversible and regenerative, and able to operate in all four quadrants of the torque/speed diagram.

5.2.7   Multiple-phase operation

To produce torque at all rotor positions the entire 360 degrees of rotation must be 'covered' by segments of rising inductance from different phases, as shown in Figure 3.18, and the phase currents must be sequenced to coincide with the appropriate segments. The total torque averaged over one revolution is usually assumed to be the sum of the torque contributions from each phase. Although the calculation and control of torque are both referred to one phase, some degree of overlap is required in practice to minimize notches in the instantaneous torque waveform when the phases are commutated, and to produce adequate starting torque at all rotor positions.

5.3   Current regulation

5.3.1   Soft chopping, hard chopping, and conduction modes

At high speed the current is controlled solely by the on/off timing of the power transistor switching, but at low and medium speeds it is regulated by chopping. This means that the power transistors are switched on/off, usually at a high frequency compared with the fundamental frequency of the phase current waveform. The voltage applied to the winding terminals is $+V_s$ if both transistors are on, 0 if one is on and the other is off, and $-V_s$ if both transistors are off and the phase current is freewheeling through both diodes. In the zero-volt state the phase current freewheels through one transistor and one diode. These three conduction modes are shown in Figure 5.9, and Table 5.1 shows the states of the power transistors and diodes in the three conduction modes.

Soft chopping is when only one transistor is chopping. The other transistor remains on, and it is called the 'commutating' transistor because its only function is to steer or commutate the current into its associated phase winding at the beginning and end of the conduction period. The voltage applied to the winding switches between $+V_s$

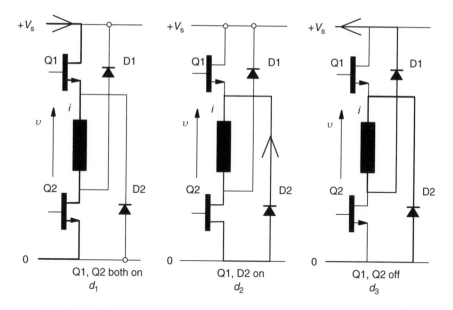

Fig. 5.9 Conduction modes.

Table 5.1 Truth table for the states of the transistors and diodes

State	Q1	Q2	D1	D2	V
A	1	1	0	0	V_s
B	1	0	1	0	0
C	0	1	0	1	0
D	0	0	1	1	$-V_s$

and 0. During the zero-volt period the rate of change of flux-linkage is very small (in fact it is equal to $-Ri$), and therefore the current falls slowly. This means that the chopping frequency and d.c. link capacitor current can both be greatly reduced for a given current ripple or hysteresis band (see below).

Hard chopping is when both transistors are switched on/off together. It generally produces more acoustic and electrical noise, and increases the current ripple and d.c. link capacitor current for a given current ripple or hysteresis band. It is necessary in certain conditions particularly during regeneration, to prevent loss of control of the current waveform, and of course the final 'chop' at θ_c at the end of the conduction period is a hard chop.

5.3.2   Single-pulse control at high speed

The flux must be established from zero every stroke, and its build-up is controlled by switching both power transistors on at the *turn-on angle* θ_0 and switching them off at the commutation angle θ_c. In motoring operation the dwell $\Delta\theta = \theta_c - \theta_0$ is timed to coincide with a period of rising inductance (rotor and stator poles of the relevant phase approaching), and in generating operation with a period of falling inductance (rotor and stator poles separating). At a sufficiently high speed, the waveforms of voltage, flux-linkage, current, and idealized inductance are as shown in Figures 5.3 and 5.4 (motoring) and Figure 5.7 (generating). The 'idealized' inductance that would be obtained with no fringing and with infinitely permeable iron has a waveform similar to that of the pole-overlap waveform, and provides a convenient means for relating the waveforms to the rotor position. Section 3.5.2 gives a simple analysis of the firing angles, which are examined in more detail in this chapter.

At constant angular velocity ω the build-up of flux-linkage proceeds according to Faraday's law:

$$\psi_c = \frac{1}{\omega} \int_{\theta_0}^{\theta_c} (V_s - Ri)\, d\theta + \psi_0 \tag{5.1}$$

where ψ_0 is the flux-linkage pre-existing at θ_0 (ordinarily zero), V_s is the supply voltage, R is the phase resistance, and i is the instantaneous current. All impedances and volt-drops in the controller and the supply are ignored at this stage. (5.1) can be written as

$$\omega\psi_c = V_s(1 - u_1) \cdot \theta_D \tag{5.2}$$

where $\theta_D = (\theta_c - \theta_0)$ is the *dwell* and $v_1 = u_1 V_s$ is the mean volt-drop in the resistance and transistors during θ_D. If $u_1 \ll 1$ the flux-linkage rises linearly. In motoring operation the flux should ideally be reduced to zero before the poles are separating, otherwise the torque changes sign and becomes a braking torque. To accomplish this the terminal voltage must be reversed at θ_c, and this is usually done by the action of the freewheeling diodes when the transistors turn off. The angle taken for the negative voltage to drive the flux back to zero at the 'extinction angle' θ_q is again governed by Faraday's law:

$$0 = \psi_c + \frac{1}{\omega} \int_{\theta_c}^{\theta_q} (-V_s - Ri)\, d\theta \tag{5.3}$$

and this can be written as

$$\omega\psi_c = V_s(1 + u_2)(\theta_q - \theta_c) \tag{5.4}$$

where $v_2 = u_2 V_s$ is the mean volt-drop in the resistance and diodes in the *de-fluxing period* $(\theta_q - \theta_c)$. If $u_2 \ll 1$ the flux-linkage falls linearly, and at constant speed the angle traversed is nearly equal to the dwell angle, both being equal to ψ_c/V_s. The peak flux-linkage ψ_c occurs at the commutation angle θ_c. The total angle of phase current conduction covers the fluxing and de-fluxing intervals and is equal to

$$\theta_q - \theta_0 \simeq \frac{\omega\psi_c}{V_s} \cdot \frac{2 - u_1 + u_2}{(1 + u_2)(1 - u_1)}. \tag{5.5}$$

If $u_1 = u_2 = 0$ this reduces to $2\omega\psi_c/V_s$. The entire conduction period must be completed within one rotor pole-pitch $\alpha_r = 2\pi/N_r$, otherwise there will be a ratcheting or pumping effect in which ψ_0 has a series of non-zero values increasing from stroke to stroke.[2] This condition is also called 'continuous conduction'. That is, $\theta_q - \theta_0 \leq \alpha_r$. Equations (5.2) and (5.5) combine to give the maximum permissible dwell angle,

$$\theta_{Dmax} = \alpha_r \cdot \frac{1 + u_2}{2 - u_1 + u_2}. \tag{5.6}$$

If the mean volt-drops u_1 and u_2 are both approximately the same fraction of V_s, so that $v_1/V_s = v_2/V_s = u$, then (5.6) reduces to

$$\theta_{Dmax} = \alpha_r \cdot \frac{(1 + u)}{2}. \tag{5.7}$$

For example, in a symmetrical 6/4 motor the pole-pitch is $\alpha_r = 90°$ (360 elec.°) and if $u = 0$ the maximum dwell angle is $\theta_D = 45°$, giving a total angle of conduction in the phase winding of 90°. But if $u = 0.2$ the maximum dwell angle is 54°. In a regular switched reluctance motor the angle of rising inductance is only $\alpha_r/2$. Ideally the flux should be zero throughout the period of falling inductance, because current flowing in that period produces a negative or braking torque. To avoid this completely, the conduction angle must be restricted to $\alpha_r/2$ and the maximum dwell angle is then

$$\theta_D \leq \frac{\alpha_r}{2} \cdot \frac{1 + u}{2}. \tag{5.8}$$

In the 6/4 motor, with $u = 0.2$ this indicates a maximum dwell angle of 27° (108 elec.°) and a conduction angle of 54°. In practice, larger dwell angles than this are used because the gain in torque-impulse during the rising-inductance period exceeds the small braking-torque impulse, which generally occurs in a region when the torque/ampere is low (i.e. near the aligned and/or unaligned positions). This condition is shown in Figures 5.3 and 5.4, where the current has a 'tail' extending beyond the aligned position. The torque is negative during this tail period, but it is small.

The turn-on angle in Figure 5.3 is just after the unaligned position, and the current rises linearly until the poles begin to overlap. The rising inductance generates a back-EMF which consumes an increasing proportion of the supply voltage, until at the peak of the current waveform the back-EMF equals V_s. Subsequently the back-EMF grows

---

[2] Note that α_r is the angle of rotation between two successive aligned positions.

greater than V_s because the flux-linkage is still increasing, while the speed is constant. What was an excess of applied forward voltage now becomes a deficit, and the current begins to decrease. At the point of commutation the applied terminal voltage reverses, and there is a sharp increase in the rate of change of current. At the aligned position the back-EMF reverses, so that instead of augmenting the negative applied terminal voltage, it diminishes it, and the rate of fall of current decreases. In this period there is a danger that the back-EMF may exceed the supply voltage and cause the current to start increasing again. It is for this reason that in single-pulse operation, commutation must precede the aligned position by several degrees. The commutation angle must be advanced as the speed increases.

Figures 5.3 and 5.4 also show the importance of switching the supply voltage on before the poles begin to overlap. This permits the current to grow to an adequate level while the inductance is still low. For as long as the inductance remains nearly constant, there is no back-EMF and the full supply voltage is available to force the increase in current. The turn-on angle may be advanced well ahead of the unaligned position at high speed, even into the previous zone of falling inductance.

5.3.3   Current regulation and voltage-PWM at low and medium speeds

The method of current regulation is a question of the timing and width of the voltage pulses. Broadly speaking there are two main methods: current-hysteresis control and voltage-PWM control, but many variations exist on these basic schemes. The drive circuit is assumed to be the same for both methods, although several variants of drive circuit have been devised to effect various improvements in the current waveform control or to reduce the cost of the controller. In both cases there is a 'flux-building' interval from initial turn-on θ_0 to commutation at θ_c, when the flux is built up from zero to its peak value. This interval is called the 'dwell' or 'transistor conduction angle'. At θ_c both transistors are switched off, and the freewheeling diodes connect the reverse of the supply voltage to the phase terminals, causing the flux to decay to zero. The 'de-fluxing interval' lasts from θ_c to θ_q, and in general is shorter than the fluxing interval.

Voltage-PWM
In voltage-PWM (Figure 5.10), at least one of the two transistors in a phaseleg is switched on and off at a predetermined frequency f_{chop}, with a duty-cycle D which is interpreted as $t_{on} \times f_{chop} = t_{on}/T_{chop}$, t_{on} being the on-time and $T_{chop} = 1/f_{chop}$ being the period at the switching frequency; see (8.49). In voltage-PWM there is no closed-loop control of the instantaneous current. The current waveform has its 'natural' shape at all speeds, as though the supply voltage was 'chopped down' to the value $D \times V_s$. However, for safety and protection a current-limiting function is provided such that if the current reaches a predetermined level i_{Hi}, the current will be limited by switching off at least one of the phaseleg transistors.

With soft chopping, Q2 remains on throughout the dwell angle. When Q1 is on, voltage V_s is connected to the phase winding. When it is off, the winding is short-circuited through Q2 and D2. Q1 is called the 'chopping transistor' and D2 the 'chopping diode'. Q2 is called the 'commutating transistor' and D1 the 'commutating

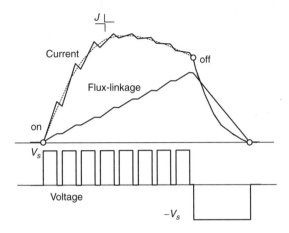

Fig. 5.10 Voltage-PWM waveforms with soft chopping.

diode', because they change state only at the commutation angles θ_0 and θ_c. During the dwell angle the average voltage applied to the phase winding is $D \times V_s$. Again using u to represent the averaged per-unit effect of volt-drops in the resistance and the semiconductors, the flux-linkage rise in the dwell period can be equated to the flux-linkage fall in the de-fluxing period to give

$$\omega\psi_c = \theta_D(D - u)V_s = (\theta_q - \theta_c)(1 + u)V_s. \tag{5.9}$$

This can be rearranged to show that the total conduction angle is

$$\theta_q - \theta_0 = \theta_D\left[\frac{1 + D}{1 + u}\right]. \tag{5.10}$$

To prevent continuous conduction, θ_D must be restricted to

$$\theta_D < \alpha_r \cdot \frac{1 + u}{1 + D}. \tag{5.11}$$

For example, in the 6/4 motor, if $u = 0.2$ and $D = 0.5$, the maximum dwell is $1.2/1.5 \times 90 = 72°$. To prevent *any* braking torque, θ_D must be restricted to

$$\theta_D < \frac{\alpha_r}{2} \cdot \frac{1 + u}{1 + D}, \tag{5.12}$$

i.e. one-half of the absolute maximum, or 36° in the example. The dwell can be increased as the duty-cycle is decreased, up to the maximum given by (5.6) or (5.8).

A similar analysis can be carried out for hard chopping, in which both transistors are switched together at high frequency. In both soft and hard chopping, the flux-linkage waveform increases in regular steps with a more-or-less constant average slope. Before the start of overlap, the average slope of the *current* waveform is also nearly constant as the linearly increasing flux is forced into a constant inductance. Thereafter, the inductance increases more or less linearly while the flux-linkage continues to rise linearly. Consequently the current tends to become constant or flat-topped. Voltage-PWM tends to produce quieter operation than current hysteresis control.

The waveforms in Figures 5.2 and 5.6 show that at low speed, when chopping is the preferred control strategy, the whole of the absolute torque zone can be used. As is evident from (5.9), the ratio of the slopes of the rising and falling parts of the flux-linkage waveform is approximately equal to D, so that with a low duty-cycle (needed to 'throttle' the voltage at low speed), the de-fluxing is accomplished in a very few degrees, permitting late commutation.

Although the pole *arcs* do not appear in any of the equations constraining the limiting values of the firing angles, they are important in determining their *optimum* values.

The duty-cycle is typically set by the outer speed and position control loops, while the firing angles can be scheduled with speed to optimize efficiency. Figure 5.11 shows the concept of average torque control with voltage-PWM, with typical voltage and current waveforms as shown in Figure 5.10. The torque demand is represented by the duty-cycle command signal D^*, which may vary as the torque demand varies, with consequent variation in the current. Because no attempt is made to control the currents instantaneously, there is no need for current sensors in individual phases. Voltage-PWM control schemes may therefore be designed using only one current sensor at the d.c. link for over-current protection.

Current hysteresis

In current hysteresis control (Figure 5.12), at least one of the two transistors in a phaseleg is switched off when the current exceeds a specified set-point value i_{Hi}. It is switched on again when the current falls below a second level $i_{Lo} = i_{Hi} - \Delta i$, where Δi is called the 'hysteresis-band'.

Current hysteresis control maintains a generally flat current waveform, as shown in Figure 5.2 or 5.6, with ripple determined by Δi and the bandwidth of the current-regulator. At high speed, the back-EMF may prevent the current from ever reaching i_{Hi}, and then the current waveform is naturally determined by the changing induct-ance and back-EMF as the rotor rotates (this is sometimes called single-pulse mode). Figure 5.12 shows the waveforms obtained with a hysteresis-type current regulator and soft chopping, in which one power transistor is switched off when $i > i_{Hi}$ and on again when $i < i_{Lo}$. The instantaneous phase current i is measured using a wide-bandwidth current transducer, and fed back to a summing junction. The error is used directly to control the states of the power transistors. Both soft and hard chopping schemes

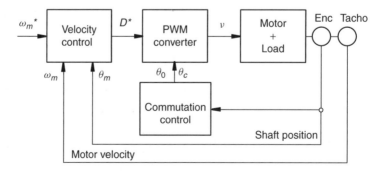

Fig. 5.11 Architecture of voltage-PWM controller.

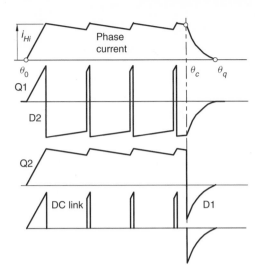

Fig. 5.12 Device current waveforms.

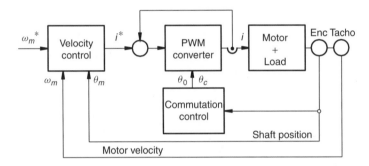

Fig. 5.13 Architecture of current-hysteresis controller.

are possible, but only the soft-chopping waveforms are shown in Figure 5.12. The waveforms for hard chopping are similar. As in the case of voltage-PWM, soft chopping decreases the current ripple and the filter requirements, but it may be necessary in braking or generating modes of operation.

Delta modulation

A variant of the current hysteresis controller is delta modulation in which the current is sampled at a fixed frequency. If the phase current has risen above the reference current i^* the phase voltage is switched off, and if it has fallen below i^* it is switched on. The switching frequency is not fixed but is limited by the sampling rate.

5.3.4 Additional current regulation techniques

Soft braking

Soft braking is used for regenerative braking and uses the same zero-voltage loop principle as soft chopping. Initially both transistors are on and full supply voltage is

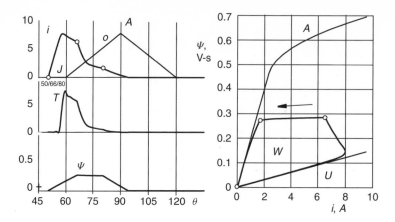

Fig. 5.14 Zero-volt loop used to achieve an interval of approximately constant flux-linkage.

applied until the phase current in the winding exceeds a predetermined limit, $i > i_{Hi}$. Both transistors are then switched off and $-V_s$ is applied until the current falls below i_{Lo}. Thereafter one transistor is switched on and off to regulate the current until the end of the conduction period. The state of the switches alternates between B and D, or between C and D, in Table 5.1. During the freewheeling state B or C, the flux-linkage remains approximately constant and at low speed the rate of rise of current is low, so that this strategy can be used to limit the switching frequency and the current ripple. At high speed, however, the current rise may be too rapid during the zero-volt periods B and C, and hard chopping may become necessary, in which the supply voltage V_s is used to suppress the rate of rise of current and the switch states alternate between A and D. Generated energy is returned to the d.c. link during state D.

Zero-volt loop

Just as current-hysteresis control can be used to maintain constant current, zero voltage can be used to maintain constant flux-linkage during part of the stroke (Miller *et al.*, 1985a). An example is shown in Figure 5.14.

By this means it is possible to reduce the torque without current chopping and its associated losses. The peak flux is limited to a lower value. The energy conversion loop still makes good utilization of the available energy. This technique has been used also for noise reduction (Horst, 1995).

5.3.5　Mathematical description of chopping

For a mathematical analysis or computer simulation it is necessary to have a definition of firing angles with respect to a reference value of rotor position, such as the one in Figure 5.15. The rotor position reference is derived from the graph of 'per-unit overlap' between the stator poles of a reference phase and any pair of rotor poles. This graph is periodic with a period of α_r, the rotor pole-pitch, and one period defines a range of 'principal values' of the rotor position θ, which may conveniently be defined to start at the 'previous aligned position' and end at the 'current aligned position', A. In a computer simulation, the rotor angle increases continuously, and when it is

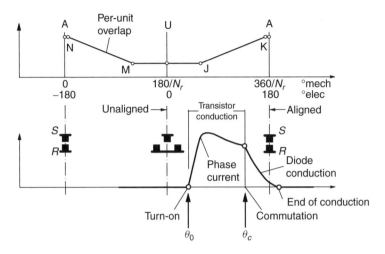

Fig. 5.15 Definition of firing angles.

outside the principal range it can be 'reduced' to the principal range simply by adding or subtracting integral multiples of the rotor pole-pitch angle α_r. In a symmetrical machine the magnetization curves are generally defined only between the unaligned and aligned positions, and it is convenient to divide the principal range into two sections AU and UA. If θ is in AU, it is replaced by $A - \theta$ so that the reduced rotor angle is always in the range UA. It is necessary to set a flag S_T to -1 to represent the sign of the torque when θ is in the AU range, and $+1$ when it is in the UA range. Otherwise all calculations can proceed as if the rotor was in the range UA. If the machine is not symmetrical, the principal range cannot be divided and must extend over a complete rotor pole-pitch; moreover, the magnetization curves must be available over this entire range.

Figure 5.15 also shows the definition of 'electrical degrees'. The origin for electrical degrees is the unaligned position and one electrical cycle is equal to the rotor pole-pitch. Therefore the conversion from electrical degrees to mechanical degrees is

$$\theta_{elec.} = (\theta_{mech.} - U) \times N_r. \tag{5.13}$$

For example, in a three-phase 6/4 motor with a pole-arc of 30°, the J position (start of overlap) is at 60° (mech.) or $(60 - 45) \times 4 = 60°$ (elec.). The interval between unaligned and aligned positions is always 180° (elec.).

Referring to the power circuit in Figure 5.9, transistors Q1 and Q2 are turned on at θ_0 and off at θ_c. In current hysteresis control, the applied voltage during the conduction interval $\theta_c - \theta_0$ is V_s. In a computer simulation, at the end of each integration step the phase current i is compared with i_{Hi}. If $i > i_{Hi}$, Q1 is switched off; otherwise it is switched on. In hard chopping, Q2 is switched off as well as Q1. When i falls below i_{Lo}, the chopping transistor is switched on again. In voltage-PWM, the chopping transistor is switched on and off at the frequency f_{chop}, with a duty-cycle D. At θ_c, Q1 and Q2 are switched off, and the reverse voltage (or 'de-fluxing' voltage) is $-V_s$. The voltage equation for one phase is:

$$\omega\frac{\mathrm{d}\psi}{\mathrm{d}\theta} = d_1[V_s - R_{ph}i - 2R_qi - 2V_q] + d_2[-R_{ph}i - R_qi - V_q - V_d]$$
$$+ d_3[-V_s - R_{ph}i - 2V_d] \tag{5.14}$$
$$= (d_1 - d_3)V_s - R_{ph}i - 2(d_1 + d_2)(R_qi + V_q) - (d_2 + 2d_3)V_d$$

where d_1 is the duty-cycle with Q1 and Q2 on, d_2 with Q1 off, and d_3 with Q1 and Q2 off. This equation caters for all combinations of the states of Q1 and Q2. At each integration step d_1, d_2 and d_3 are assigned the correct values, and $d_1 + d_2 + d_3 = 1$ within each integration step. The compact voltage equation expressed in this form with d_1, d_2 and d_3 embodies all the switching states and logic required for both current hysteresis control and voltage-PWM control. In current hysteresis control, d_1, d_2 and d_3 are scalar values that multiply the voltage terms in the equation. This is the principle of 'state space averaging', and is based on the notion of an upstream chopper controlling the d.c. source voltage, with infinite chopping frequency. d_1 is either equal to D or 0; d_2 is either $1 - D$ or 0; and d_3 is either 0 or 1. In voltage-PWM, d_1, d_2 and d_3 are binary states having the value 0 or 1. The states are determined by the combined states of the transistors and diodes. Only one of the three states d_1, d_2 and d_3 can be non-zero during one integration step.

The transistor and diode currents and their squares are accumulated in each integration step using $i_{Q1} = d_1 \times i$; $i_{Q2} = (d_1 + d_2) \times i$; $i_{D1} = d_3 \times i$; $i_{D2} = (d_2 + d_3) \times i$; and $i_{DC} = (d_1 - d_3) \times i\{= i_{Q1} - i_{D1} = i_{Q2} - i_{D2}\}$. When the integration is finished, the mean and mean-squared values are calculated from the accumulations by dividing by the number of steps. This process is the same for both current hysteresis control and voltage-PWM. An exception to this calculation is the d.c. link ripple current. This has to be constructed from the phase-shifted sums of the phaseleg currents, which flow in both directions in the d.c. link. It can be constructed from the array of samples of phase current.

5.4 Regulation algorithms

5.4.1 Motor control

As already noted at the beginning of this chapter, the regulation algorithm in d.c. motor drives is based on the simple linear relationship between torque and current, and the torque demand produced by the velocity loop is translated directly into a current command by a simple constant of proportionality, the torque constant k_T. A similar principle is implemented in field-oriented a.c. motor drives. The translation of the torque demand signal T^* into a current command signal i^* is the function of the 'feedforward' part of the torque controller, since there is usually no torque transducer and therefore no torque feedback.

In the switched reluctance drive the relationship between torque and current is not linear. An example is shown in Figure 5.16.

The torque also depends on the firing angles θ_0 and θ_c. To complicate matters, θ_0 and θ_c may be required to vary for reasons other than torque control – for example, to minimize noise or to compensate for back-EMF at high speed. The result is that the feedforward torque control must usually be implemented as a mapping from T^* to i^*, with additional links to the firing angles and possibly also the supply voltage. The mapping can be implemented in a digital memory, with interpolation in the processor,

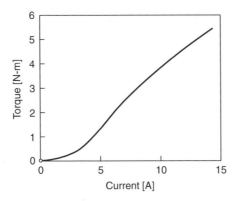

Fig. 5.16 Typical variation of torque with current.

or possibly by equations. In either case, the mapping must be computed or determined by experiment; this can be a laborious and time-consuming process. Unfortunately the mapping will be specific to a particular motor and drive, and usually also specific to a particular application.

Figure 5.17 shows a graphical representation of a look-up table for the turn-on angle θ_0 as a function of speed and torque demand. With single-pulse control the variation of torque with θ_0 and θ_c is equally complex. The architecture of a single-pulse controller is shown in Figure 5.18.

For closed-loop control of the phase currents both linear and nonlinear current regulators may be used (Kjaer *et al.*, 1996). Linear regulators normally use proportional-integral (PI) control to eliminate the error between the reference and actual currents, and to give a smooth variation of phase voltage with reduced torque ripple and electromagnetic noise. The main disadvantage stems from the variation of inductance with rotor position, which can cause the electrical time constant to vary by as much as 10:1, making it difficult to tune for satisfactory transient performance (see Chapter 9).

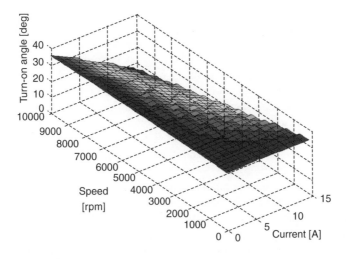

Fig. 5.17 Graphical representation of look-up table for turn-on angle.

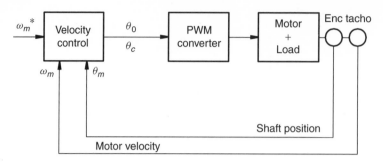

Fig. 5.18 Architecture of single-pulse controller.

5.4.2 Generator control

The switched reluctance machine will regenerate power to the d.c. supply if the current pulse is timed to coincide with an interval of falling inductance, and typical generator waveforms are shown in Figures 5.6 and 5.7. Excitation power is supplied by the d.c. source when the transistors are both on, and generated power is returned to the d.c. source when they are both switched off. The power circuits of Figures 3.7 and 3.8 may be used, but several variants have been published, (e.g. Radun, 1994).

In the steady state the switched reluctance machine can sustain itself in the generating mode with the d.c. source disconnected, but the d.c. link capacitor must be retained to provide excitation power during the 'fluxing interval' during the first part of each stroke. Generally the load will be connected in parallel with the d.c. link capacitor, and in general its impedance will be variable and not under the control of the switched reluctance generator controller. Inevitably the d.c. voltage will decrease during the fluxing interval, and increase during the de-fluxing interval, which is when power is being returned through the diodes. The variation or ripple in the d.c. link (capacitor) voltage depends on the energy conversion per stroke, the energy ratio, and the capacitance. The controller must maintain the average d.c. voltage constant in much the same way as it must maintain constant average torque in motoring mode. Indeed, if the average d.c. voltage is constant and the speed is constant, then if losses are neglected, the maintenance of constant d.c. current is equivalent to the maintenance of constant torque, both being as averaged over at least one stroke. Therefore, in principle, the architecture of a generator controller is similar to that of a motor controller. However, the d.c. link capacitance has an integrating or smoothing effect such that it requires lower bandwidth to control the d.c. voltage than to control the d.c. current (or torque). At low speed, therefore, the d.c. link voltage is controlled by varying the set-point current i^* or the duty-cycle D^*, and at high speed by varying the control angles θ_0 and θ_c (see Figure 5.19).

As in the case of motoring operation, a control map is required to determine how the control variables must vary in response to the voltage error ΔV_d. Various linearizing schemes have been presented to simplify the control laws (Radun, 1993; Kjaer *et al.*, 1994) (see Chapter 10).

Although the d.c. source is necessary during startup from 'cold', i.e. with the machine unmagnetized and no voltage on the d.c. link capacitor, it can be disconnected once the

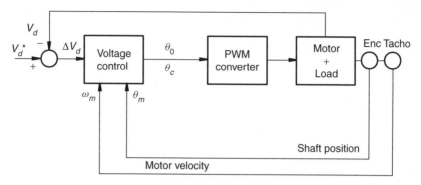

Fig. 5.19 Architecture of generator controller (high-speed mode).

system is self-sustaining. Whether or not this is done is a matter of design philosophy for the particular application.

5.4.3 Optimization of the control variables

Table 5.2 summarizes the different control schemes for torque (in motoring operation) or d.c. voltage (in generating operation), and their associated control variables.

Average torque can be controlled by varying any one or indeed all of the control variables in a given mode, but the configuration depends on the performance requirements, the acceptable level of complexity, and the cost. Since there are many possible combinations of control variables which produce the same torque, a secondary control objective is needed to select and define the variation of control variables for optimum performance. Examples of such secondary control objectives are efficiency, acoustic noise, and torque ripple. Obviously the nonlinearity of the switched reluctance machine can make the task complicated, and it is usually necessary to undertake extensive simulations and/or dynamometer tests. Attempts have been made to define set procedures for calculating these parameters (Orthmann and Schoner, 1993; Stiebler and Ge, 1992; Takahashi *et al.*, 1993; Torrey and Lang, 1990; Gribble *et al.*, 1996; Gribble *et al.*, 1999; Kjaer *et al.*, 1995).

Some early optimization methods were valid for fixed speed and load, and in some cases they required extensive simulations of the machine model and a detailed system knowledge. More recent generic solutions can be applied to different motors and remain valid under a change in operating conditions. An example of a generic efficiency optimization scheme for a current-controlled motor is in Kjaer *et al.*, (1995). In this scheme, the current reference variable and the turn-on angle are defined to maximize the average torque while the turn-off angle is derived to maximize efficiency. The

Table 5.2 Control modes with their control variables

Control mode	Control variables
Current hysteresis control	i_{Hi}, i_{Lo}
Delta-modulation	i^*
Voltage-PWM control	D
Zero-volt-loop mode	i_{Hi}, i_{Lo}, θ_z
Single-pulse control	θ_0, θ_c

current demand is set by the velocity loop. The turn-on angle θ_0 is varied as a function of current demand i^* according to the equation

$$\theta_0 = \theta_J - \frac{\omega_m L_u i^*}{V_s} \qquad (5.15)$$

(Bose, 1985), so that the phase current reaches its reference value at the point where the phase inductance is beginning to increase, i.e. at the J position where torque production starts. Under transient conditions the turn-off angle θ_c is set as a function of motor speed; but in the steady state, when the outer loop is stable and the current reference is constant, it is adjusted to maximize efficiency. The efficiency is given by

$$\eta = \frac{T\omega_m}{V_s I_{d.c.}} \qquad (5.16)$$

where $I_{d.c.}$ is the mean d.c. supply current. With a slowly varying load, at constant speed and fixed d.c. link voltage, the efficiency is optimized by minimizing the d.c. link current, $I_{d.c.}$. A control loop, considerably slower than the outer speed control loop, is used to adjust θ_c to minimize $I_{d.c.}$. Since the efficiency-optimizing loop is much slower than the speed control loop, the dynamic performance is not affected.

Efficiency optimization techniques have also been developed for switched reluctance generators (Kjaer *et al.*, 1994). The machine is assumed to be operating in single-pulse mode with closed-loop control of the d.c. link voltage. An 'inverse' model is used to cancel the nonlinearities of the machine so that the average generated current can be represented as function of the d.c. link voltage, the speed, and the firing angles (Kjaer *et al.*, 1994) (see Chapter 10).

Sometimes it is not feasible to vary the firing angles on-line because the transient response might be compromised. 'Off-line' optimization methods have therefore been developed to maximize the energy conversion loop in both voltage and current control modes (Gribble *et al.*, 1996; Gribble *et al.*, 1999). Secondary objectives are to maximize efficiency under current control, or to produce approximately flat-topped current waveforms under voltage control. The model used to calculate θ_c is kept simple and generic, to make it applicable to different machines and easy to calculate.

Other optimization objectives include the minimization of torque ripple and the reduction of acoustic noise. For applications which require optimization of multiple performance criteria, it is virtually essential to run a series of dynamometer tests to obtain the requisite control laws (see Chapter 9).

5.5 Summary

This chapter has introduced the basic strategies for controlling the average torque and the current waveform over the whole range of speed and torque, in all four quadrants under normal operating conditions. The focus has been on strategies which use the minimum number of control variables, for example a turn-on angle, a turn-off angle, and a reference current. These strategies are capable of controlling the average torque, but the variation of torque through one stroke is uncontrolled and is determined by the magnetization characteristics of the machine (see Chapter 6). The shape of the energy-conversion loop is an important guide to the effectiveness of the basic control strategies, and it is often the main objective in average torque control to maximize its area for a given peak or r.m.s. current. The basic control strategies divide into two main groups at low speed: those which regulate the current and those which modulate the voltage by pulse-width modulation. At high speed, when the back-EMF of the machine is high, both strategies merge into *single-pulse* control, in which there is no chopping, but only the firing angles are used for control. The concept of the zero-volt loop has been described as an additional control means, which can help to shape the energy-conversion loop at any speed. Several examples are included showing typical waveforms of current, torque, and flux-linkage, and energy-conversion loops at different speeds and torques.

6

Instantaneous torque control

Philip C. Kjaer
ABB Corporate Research, Västerås, Sweden

and

Jeremy J. Gribble
DERA, Pyestock, UK[1]

6.1 Introduction

This chapter describes the development of an advanced control scheme that permits high-grade control of the switched reluctance motor. 'High-grade' in this context means performance that will enable the switched reluctance machine to be considered for applications currently employing vector controlled a.c. drives, brushless d.c., or even d.c. commutator motor drives. The specific features of high-grade control include: (i) high-bandwidth torque control; (ii) low torque ripple; (iii) four-quadrant operation over a wide speed range including zero speed; and (iv) seamless transition from maximum-efficiency operation to maximum field-weakening with efficient use of the controller voltage. This control strategy is regarded as the functional equivalent of a.c. vector control.

The motor model is based on the assumptions that (i) there is no mutual magnetic coupling between phases; (ii) iron (and other) losses and phase resistance are negligible; and (iii) all electromagnetic parameters can be calculated analytically or numerically. It is assumed that the m-phase machine is fed from a classic asymmetric, lossless converter with $2m$ switches. The angular displacement in electrical degrees between neighbouring phases $j - 1$ and j is defined as:

$$\theta_{pitch} = \frac{360°}{N_r}$$

$$\theta_j = \theta_{j-1} - \theta_{pitch}. \qquad (6.1)$$

---

[1] The work in this chapter was completed in 1997 at Glasgow University.

By superposition, the total torque is expressed as the sum of the m individual phase torques:

$$\tau_{total} = \sum_{j=1}^{m} \tau_j = \sum_{j=1}^{m} \tau(i_j, \theta_j).$$ (6.2)

To achieve the high-bandwidth torque control needed in demanding applications requiring good transient performance, it is necessary to employ instantaneous torque control, and to replace the notion of plain commutation with a control scheme that carefully profiles the phase currents to produce the desired total motor torque by co-ordinating the torques produced by individual phases.

6.1.1   Formulation of torque control problem

If the individual phase currents can be controlled with a high bandwidth, then the problem of choosing the appropriate current waveforms to ensure low torque ripple becomes the key issue. Put another way, the production of torque by individual phases must be co-ordinated so that the total torque tracks the reference value generated by the position or speed control loops. Unfortunately there is no inherent unique distribution of torque between phases, so the torque-sharing functions and the associated current or flux waveforms must be designed. Fixing the torque waveform to some *ad hoc* function is not satisfactory because the associated current and flux-linkage waveforms directly affect the ohmic losses and the feeding voltage required (Ilic-Spong *et al.*, 1987a). To select an optimal torque-sharing function whose primary objective is low ripple torque control, it is necessary to specify a secondary objective.

6.2   Previous work and review

The first reported attempt to reduce the torque ripple was the four-phase switched reluctance servo drive presented by Byrne *et al.* (1985). The machine was designed to give a sinusoidal torque-per-phase for a constant current level. When excited with sinusoidal half-wave currents two overlapping phases could produce constant total torque, with torque ripple reported to be as low as 5%. It was recognized by Ilic-Spong *et al.* (1987a) that the total torque in general could be kept ripple-free by sharing it between individual phases, and the term 'torque sharing functions' (TSF) was introduced. The switched reluctance machine differs from the conventional three-phase a.c. machine in that the total torque may be shared more or less arbitrarily between phases, as they are largely independent.

The nonlinear torque problem has received special interest from control engineers, for example Ilic-Spong *et al.* (1987b), where the feedback linearization control technique was applied to 'linearize' the switched reluctance motor. Unfortunately, commutation was neglected. Although the nonlinearity is not of a dynamic nature, for a period the switched reluctance motor was a popular 'problem' for control engineers, and works following Ilic-Spong *et al.* (1987b) can be found in Cailleux *et al.* (1993), Panda and Dash (1996), Rossi and Tonielli (1994), and Amor *et al.* (1993a,b; 1995). However, these nonlinear control techniques require analytical electromagnetic models, whereas the references cited use rather simple and inaccurate models.

There have been many attempts to fix the torque-per-phase waveforms to some *ad hoc* function (Ilic-Spong *et al.*, 1987a; Husain and Ehasani, 1994a; Hung, 1993; and Goldenberg *et al.*, 1994). Examples are Schramm *et al.* (1992), Reay *et al.* (1993a,b), where the torque-per-phase was chosen to be trapezoidal, and the proportions of the trapezia were optimized to reduce the peak phase current. However, there is little justification for a trapezoidal phase torque profile. Furthermore, this torque waveform will impose unachievable requirements on the power electronic converter, as is clear from the following example, which assumes a nonsaturable motor:

$$\frac{d\tau}{d\theta} = \frac{\partial \tau}{\partial i} \cdot \frac{di}{d\theta} = \frac{\partial}{\partial i} \left(\frac{1}{2} \cdot \frac{dL}{d\theta} \cdot i^2 \right) \cdot \frac{di}{d\theta} = \frac{dL}{d\theta} \cdot i \cdot \frac{di}{d\theta}. \tag{6.3}$$

At zero current a nonzero $d\tau/d\theta$ will require infinite $di/d\theta$.

Most of the proposed torque-sharing functions meet the primary objective of low torque ripple, but do not recognize the need for a secondary objective to determine the torque waveforms uniquely. As any phase torque waveform can be translated into current or flux-linkage waveforms, the choice of TSF directly affects the ohmic losses (ri^2) and the feeding voltage ($d\psi/dt$) required to 'track' the TSF. Both issues were recognized in Ilic-Spong, (1987a), but not analysed in detail. The ohmic loss relates to the drive efficiency, and the required feeding voltage relates to the torque/speed capability. More explicitly, two possible optimization criteria may be formulated as the search for:

(a) the torque-sharing function $\tau(\theta)$ that minimizes $\int ri^2(\theta) \, d\theta$;
(b) the torque-sharing function $\tau(\theta)$ that minimizes $|v(\theta)|_{max}$;

or even a torque-sharing function that combines the two. Similar optimization objectives were formulated for brushless PM a.c. motor drives (Favre *et al.* 1993) and for average torque control in Chapter 5. For instantaneous torque control, the low-loss TSF (a) and the low-voltage TSF (b) can be found using numerical minimization methods, though this represents a demanding computational task.

Filicori (1993) and Rossi and Tonielli (1994) presented torque-sharing waveforms derived by nonlinear optimization techniques for a three-phase low-speed high-torque direct-drive motor. A TSF that minimized the ohmic losses and one that minimized the phase voltage were presented. Measured magnetization data were fitted to a nonlinear flux-linkage model involving Fourier series. A simplex optimization algorithm (Press, 1995) was applied with the two cost functions (a) and (b). The low-loss TSF was derived at zero speed without constraints on the phase voltage, and as expected the available voltage was insufficient to track the TSF at higher speeds. The low-voltage optimization resulted in a TSF where each phase should conduct continuously – even when the phase produces negative torque. This is very different from the traditional 'switched' quasi-rectangular current waveforms. The reason for continuous conduction is obviously that the τ/ψ ratio increases with the flux level, and therefore so does the $\Delta\tau/\Delta v$ ratio. However, continuous conduction implies unnecessarily large copper losses. Minimum voltage itself is not an objective – rather it is of interest that the required phase voltage never exceeds the available controller voltage: $|v(\theta)|_{max} \leq V_{DC}$.

Though Filicori (1993) presented torque-sharing waveforms that mathematically optimized the two criteria given, the waveforms are of restricted practical use, and severe computational difficulties are associated with calculating the waveforms.

6.2.1   Choice of phase and pole-numbers

A more promising approach to low ripple operation was presented in Wallace and Taylor (1991, 1992), which have some similarity with the work presented here. The details of these will be covered later. In Wallace and Taylor (1991) the choice of motor geometry when operating with low ripple waveforms was discussed. Two-phase switched reluctance motors can be designed such that torque can be produced at any angular position, but these 'irregular' designs are restricted to unidirectional operation (Miller, 1993). Designs with a phase number of three or more can provide torque in both directions, and can therefore start from any angular position. Hence, for a high-performance four-quadrant switched reluctance motor drive, at least three phases are required. In Wallace and Taylor (1991) four three-phase designs (two 6/4-pole and two 18/20-pole motors) were compared by simulation. For both pole configurations the ratio between pole arc and pitch was varied (1/3 and 1/2), to see the effect on low ripple operation. A low arc/pitch ratio results in high saliency and high peak torque (normally maximum at the rotor positions where the poles start to overlap), but the torque variation with respect to position for constant current becomes 'flatter' when the arc/pitch ratio is increased. The findings in Wallace and Taylor (1991) indicated that high saliency in low pole-number designs increased the available low-ripple torque for a given electrical loading, whereas the opposite observation was made for the high pole-number machine. In general, three-phase motors designed for maximum output (saliency) will, at constant excitation current, exhibit large zones of low torque and to avoid the 'torque-dips' the current must be boosted at these positions (Miller, 1993). The phase currents can be excessive for this kind of operation, and it is generally accepted that at least four phases are preferred for high-performance switched reluctance motor drives to reduce the 'inherent' torque variations. Higher phase numbers can be used, but with increased complexity and commutation frequencies, and often lower efficiency. Control and design of the switched reluctance motor are strongly connected, but it has not been attempted to suggest design improvements based on the new control strategies presented in this work. It should be stressed that this work is not confined to low-speed machines.

6.3   Theory of instantaneous torque control

6.3.1   Maximum torque per ampere control

Numerical electromagnetic models and powerful computer-based optimization algorithms were used in Finch et al. (1986) without, and in Lovatt and Stephenson (1994) with limited converter VA-ratings to determine the waveforms that produce maximum average torque per r.m.s. current. Both found that the optimum current waveforms resembled the static torque-position curves, that the required driving voltage was excessive, and that the torque ripple was significantly increased when compared to square-wave excitation. The cumbersome and slow process of deriving these optimum-efficiency waveforms becomes even worse when adding the constraint that the sum of the phase torques must remain constant. Exhaustive search algorithms, simplex gradient routines (Press, 1995) or dynamic programming are required for this difficult task. Effectively one attempts to optimize the phase torque at every angle with an objective that relies on the entire waveform. Instead of minimizing the r.m.s. current, a similar objective may be formulated: minimize the phase current at all angles for a given torque. This is an objective which applies to each individual rotor position alone and does not rely on the entire waveform; it is representative of maximum efficiency operation and renders the problem tractable.

At zero speed this objective requires that only the phase with the largest torque-per-amp ratio should produce torque. This means that only one phase conducts at any time, and commutation is instantaneous. Figure 6.1 shows how the phase currents and torques would look at zero speed.

Critical rotor positions θ_c^i are defined as where two neighbouring phases can both produce the required torque (τ_d) at the same current level (equal 'strength'):

$$i(\tau_d, \theta_c^i) = i(\tau_d, \theta_c^i + \theta_{pitch}). \tag{6.4}$$

At zero speed two phases could commutate instantly at θ_c^i and maximum efficiency operation would be assured. However, at nonzero speeds instantaneous commutation is no longer possible due to the finite controller voltage. Wallace and Taylor (1992) proposed instead to define the critical rotor positions as where two neighbouring phases could produce half the required torque ($\tau_d/2$) at equal current levels:

$$i\left(\frac{\tau_d}{2}, \theta_c^i\right) = i\left(\frac{\tau_d}{2}, \theta_c^i + \theta_{pitch}\right). \tag{6.5}$$

This work uses the same notion of critical rotor positions to develop a general switched reluctance motor torque control scheme meeting both requirements of low copper losses and low voltage requirements.

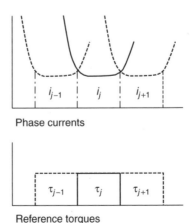

Phase currents

Reference torques

Fig. 6.1 Phase currents required to produce a constant level of torque, when only one phase conducts at a time (motor #1). (For motor data see Kjaer (1997a).)

For $\theta < \theta_c^i$ phase $j - 1$ is the stronger phase and for $\theta > \theta_c^i$ phase j becomes stronger. As it is desired that the strongest phase be the main torque contributor, the flux in other phases must be driven to, or from, zero as rapidly as possible. Commutation between phases takes place as follows: when θ approaches θ_c^i the on-coming phase j is excited with maximum available voltage so that it reaches the required torque level $\tau_d/2$ exactly at θ_c^i. During this period the flux in the retreating phase $j - 1$ is adjusted to maintain the total torque at the desired value. As θ moves past θ_c^i, phase $j - 1$ is no longer the main torque producer and it is demagnetized as quickly as possible, while the current in phase j, which is now the stronger, is adjusted to regulate the total torque (Figure 6.2). More details will be given in the following section. The magnetization/demagnetization periods are minimized through full utilization of the inverter voltage, $d\psi/d\theta = \pm V_{DC}/\omega$, unlike the approach reported in Wallace and Taylor (1992) where commutation at θ_c^i with a fixed ratio of $di/d\theta$ was performed. As discussed above, fixing the phase torque profile or the phase current profile is not optimal. The penalty paid for the more effective inverter utilization proposed here is that the torque-per-phase waveforms must change with speed, but it will improve efficiency.

Two examples may help illustrate and justify the proposed operation. If two phases (j and $j + 1$) can produce torque in the same direction at a given rotor position, how should they share the total torque in order to maximize the ratio of torque per r.m.s. current while the sum of the two phase torques is constant?

Example 1: The ideally saturable machine has a phase torque proportional to the current: $\tau_j = k_{T,j} \cdot i_j$. Let $k_0 = k_{T,j}/k_{T,j+1}$ be the ratio between the torque constants of

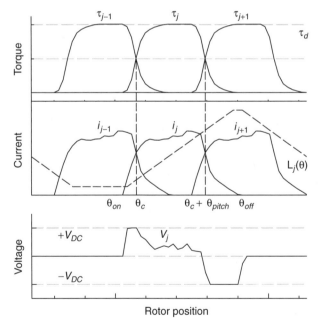

Fig. 6.2 Example of phase torque, current and voltage associated with high-efficiency torque control (motor #1), with idealized inductance profile $L(\theta)$ shown.

two phases ($k_0 \geq 1$). With the smooth torque requirement, $\tau_d = \tau_j + \tau_{j+1}$ the copper losses may be represented by:

$$I_{r.m.s.}^2 = i_j^2 + i_{j+1}^2 = \frac{1}{k_{T,j}^2} \cdot [\tau_j^2(1 + k_0^2) - 2k_0^2\tau_d\tau_j]. \tag{6.6}$$

Maximum torque per r.m.s. current is found by setting $\partial I_{r.m.s.}^2 / \partial \tau_j = 0$ which leads to:

$$\frac{\tau_j}{\tau_d} = \frac{k_0^2}{1 + k_0^2}. \tag{6.7}$$

Example 2: For the nonsaturable machine the torque per phase is $\tau_j = k_{T,j} \cdot i_j^2$ and with $k_0 = k_{T,j}/k_{T,j+1}$ the copper losses are proportional to:

$$I_{r.m.s.}^2 = i_j^2 + i_{j+1}^2 = \frac{1}{k_{T,j}} \cdot [\tau_j(1 - k_0) + k_0\tau_d] \tag{6.8}$$

which is minimized when $\tau_j = \tau_d$ and $\tau_{j+1} = 0$ for ($k_0 \geq 1$).

As the practical switched reluctance motor falls somewhere between the two examples, the results indicate boundaries for operation with maximum torque per r.m.s. current. If two phases can produce equal torques at equal currents ($k_0 = 1$), both examples justify that they should share the torque equally. If one phase is stronger than the other ($k_0 > 1$) then example 1 suggests a particular sharing function, whereas example 2 suggests that only the stronger phase conducts. Based on engineering intuition more than mathematical proof, magnetization and demagnetization with full inverter voltage, as suggested above, attempts to do exactly that.

6.3.2 Maximum torque per flux control

As described, θ_c^i represents the most efficient operating point. Eventually the available voltage will be insufficient to track the reference waveforms. To achieve another objective, namely a wide torque/speed range with low torque ripple, the required phase voltage must be kept low. Low torque ripple with low levels of $d\psi/dt$ can be achieved by keeping the flux-linkage itself low.

The electromagnetics of the switched reluctance motor imply that the flux-linkages are unequal in the two commutating phases:

$$\psi\left(\frac{\tau_d}{2}, \theta_c^i\right) < \psi\left(\frac{\tau_d}{2}, \theta_c^i + \theta_{pitch}\right). \tag{6.9}$$

To keep the flux-linkage level low, it appears preferable to commutate around another rotor position θ_c^ψ where two phases produce equal levels of torque at equal levels of flux-linkage, Figure 6.3.

In analogy with the high-efficiency operating point we get:

$$\psi\left(\frac{\tau_d}{2}, \theta_c^\psi\right) = \psi\left(\frac{\tau_d}{2}, \theta_c^\psi + \theta_{pitch}\right). \tag{6.10}$$

Hence, θ_c^ψ is believed to represent the operating point with minimum requirements on the phase voltage. Commutation and torque control is performed exactly as for θ_c^i. At

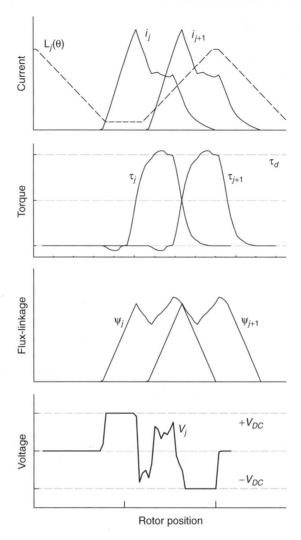

Fig. 6.3 Example of phase torque, current and voltage associated with low-flux torque control (motor #1), with idealized inductance profile $L(\theta)$ shown.

θ_c^ψ the phase currents will not be equal and therefore not kept small, and operation at θ_c^ψ will never be as efficient as at θ_c^i, as is implied by the fact that:

$$i\left(\frac{\tau_d}{2}, \theta_c^\psi\right) > i\left(\frac{\tau_d}{2}, \theta_c^\psi + \theta_{pitch}\right). \tag{6.11}$$

Hence, the theory suggests that θ_c^i and θ_c^ψ signify the two extremes of the operating range, and that the angle θ_c should be a parameter varied in between according to the operating speed, like the flux-current angle in a.c. drives. This will be verified by simulations and measurements.

6.3.3   Detailed operation of torque-sharing functions

The torque-sharing waveforms can be divided into three distinct sections.

Magnetization period: To fully utilize the available controller voltage, an 'on-coming' phase is turned on at $\theta = \theta_{on}$ and the entire inverter d.c.-link voltage V_{DC} is applied to the phase until $\theta = \theta_c$. This builds up the excitation in the phase in the shortest possible time, to assure $\tau(\theta_c) = \tau_d/2$. Therefore:

$$\theta_{on} = \theta_c - \frac{\omega \cdot \psi\left(\frac{\tau_d}{2}, \theta_c\right)}{V_{DC}}. \tag{6.12}$$

The torque and flux-linkage of phase j in this range of positions $\theta \in [\theta_{on}; \theta_c]$ are:

$$\psi_j(\theta) = \frac{V_{DC}}{\omega} \cdot (\theta - \theta_{on})$$

$$\tau_j(\theta) = \tau(\psi_j(\theta), \theta) \tag{6.13}$$

and phase $j-1$ must produce the majority of torque in this period.

Centre period: From θ_c to $\theta_c + \theta_{pitch}$ phase j is the strongest, and it carries the total torque less whatever other conducting phases might produce. This part of the waveform also changes shape with speed, but never its angular width (equal to θ_{pitch}). The torque and flux-linkage in this range of positions are:

$$\tau_j(\theta) = \tau_d - \tau_{j-1}(\theta) - \tau_{j+1}(\theta)$$

$$\psi_j(\theta) = \psi(\tau_j(\theta), \theta) \tag{6.14}$$

where τ_{j-1} and τ_{j+1} may be nonzero or not, depending on operating speed and total torque level.

Demagnetization period: Once phase j ceases to be the strongest it is turned off as quickly as possible, i.e. by applying $-V_{DC}$ to the phase until completely demagnetized, $\psi_j = 0$, which happens at:

$$\theta_{off} = \theta_c + \frac{\omega \cdot \psi\left(\frac{\tau_d}{2}, \theta_c\right)}{V_{DC}} + \theta_{pitch}. \tag{6.15}$$

The torque and flux-linkage in this range of positions $\theta \in [\theta_c + \theta_{pitch}; \theta_{off}]$ are therefore:

$$\psi_j(\theta) = \psi\left(\frac{\tau_d}{2}, \theta_c + \theta_{pitch}\right) - \frac{V_{DC}}{\omega} \cdot (\theta - \theta_c - \theta_{pitch})$$

$$\tau_j(\theta) = \tau(\psi_j(\theta), \theta). \tag{6.16}$$

6.3.4   Operating limits imposed by controller voltage

As the speed increases, so does the conduction width of the phases. It follows from the previous section that the magnetization and demagnetization periods will depend very much on θ_c. Bearing in mind, from (6.9), that the magnetization period is shorter than the demagnetization period for low-loss operation, whereas the two periods are exactly equal for low-flux operation:

$$\theta_c^i - \theta_{on} < \theta_{off} - \theta_c^i - \theta_{pitch}$$

$$\theta_c^\psi - \theta_{on} = \theta_{off} - \theta_c^\psi - \theta_{pitch} \tag{6.17}$$

the condition for which there will be periods with three phases conducting simultaneously is:

$$(\theta_c - \theta_{on}) + (\theta_{off} - \theta_c - \theta_{pitch}) \geq \theta_{pitch}. \tag{6.18}$$

The assumption that two neighbouring phases share the total torque and all other phases are driven to zero only holds up to a certain speed (albeit a very high one). This condition can be written as:

$$(\theta_c - \theta_{on}) \leq \theta_{pitch}$$

$$(\theta_{off} - \theta_c - \theta_{pitch}) \leq \theta_{pitch}. \tag{6.19}$$

Hence, for a three-phase motor the maximum permissible conduction period is 360° (elec.), and for a four-phase motor it is 270°. (6.19) can instead be formulated as a constraint on the flux-linkage level at θ_c:

$$\omega \cdot \psi \left(\frac{\tau_d}{2}, \theta_c \right) \leq V_{DC} \cdot \theta_{pitch}. \tag{6.20}$$

(6.20) together with (6.9) and (6.10) imply that for a given torque level operation at $\theta_c = \theta_c^\psi$ is possible beyond $\theta_c = \theta_c^i$. Figure 6.4 shows an example of how the current waveforms change with speed (for a four-phase motor), until an invalid operation point is reached.

As shown, the magnetization and demagnetization angles are modified according to the specific operating speed and inverter voltage. This is not the case for the centre part of the conduction period from θ_c to $\theta_c + \theta_{pitch}$, during which the torque-sharing waveform is equally likely to 'run out of volts'. Figure 6.5 shows a simulated example where the inverter voltage is insufficient to track the smooth torque waveform. Referring to (6.9) and (6.10), it would be expected that the closer θ_c is operated to θ_c^ψ the smaller the requirements to the phase voltage in the centre part would be. This seems to be the case as indicated by simulations and measurements.

For a given torque level, speed and inverter voltage, a series of values of θ_c were used. In each case the greater of flux-linkage levels at θ_c and $\theta_c + \theta_{pitch}$ and the r.m.s. phase current were calculated. The results are shown in Figure 6.6. Clearly, minimum r.m.s. current is achieved at θ_c^i and minimum 'commutating' flux-linkage is achieved at θ_c^ψ. Figure 6.8 shows simulation results of how the TSF waveforms change with θ_c.

Properties at critical angles: To complicate matters further, the locations of θ_c^i and θ_c^ψ depend on the saturation level which in turn depends on τ_d. Figure 6.7 shows simulations of how the critical angles vary with the torque level. As expected, θ_c^i is more affected than θ_c^ψ. Returning to Figure 6.6 it is clear that the r.m.s. current does not change very much around θ_c^i and it would be acceptable to make both θ_c^i and θ_c^ψ independent of τ_d in practical implementation.

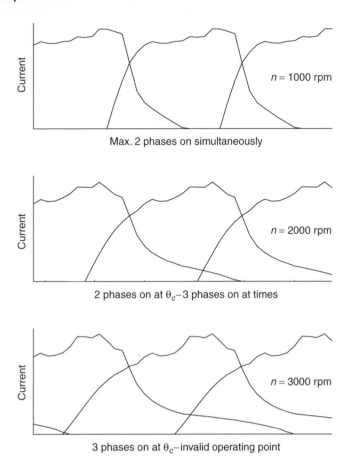

Fig. 6.4 Phase currents changing with speed for low-loss, low torque ripple operation simulated for motor #1.

With the torque-sharing waveforms determined by relatively few drive parameters and the motor electromagnetics, it would be desirable to establish analytical relationships, from which the drive performance could be estimated (see Chiba and Fukao (1992) and Betz *et al.* (1993) for a similar analysis on the synchronous reluctance drive). However, no analytical models published to date are accurate *and* simple enough to use for this task.

6.3.5   Simulated torque/speed characteristics

The torque/speed capability of the proposed low-ripple control strategy is limited by two constraints: controller voltage and motor r.m.s. current. Note that in this analysis the peak controller current has not been constrained, as there is no simple relation between r.m.s. phase current and inverter current. Numerous simulations have been performed, and for each operating point it is necessary to assure that $|v(\theta)| \leq V_{DC}$ and $I_{r.m.s.} \leq I_{r.m.s.,max}$. This meant in practice finding the maximum torque level that could be produced before the voltage waveform started to look like Figure 6.5 (insufficient

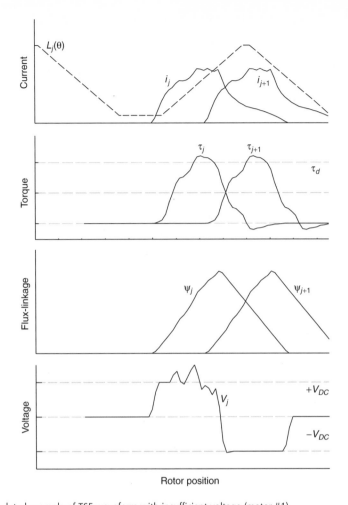

Fig. 6.5 Simulated example of TSF waveform with insufficient voltage (motor #1).

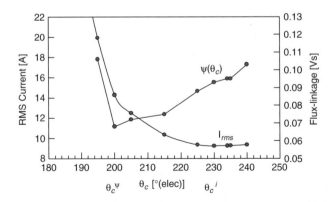

Fig. 6.6 Simulated flux-linkage $\psi(\theta_c)$ and r.m.s. current vs. θ_c, for $\tau_d = 3.5$ Nm, $\omega = 105$ rad/s, $V_{DC} = 100$ V, for motor #1.

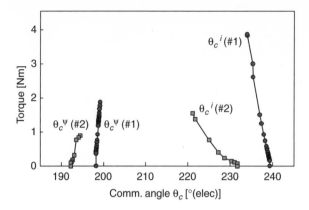

Fig. 6.7 Variation of θ_c^i and θ_c^ψ vs. τ_d for motors #1 and #2.

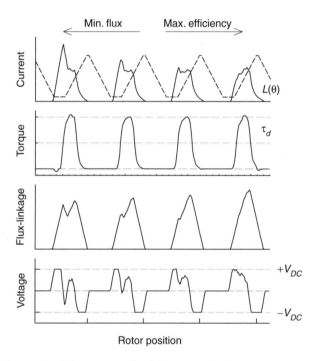

Fig. 6.8 Simulated examples of phase current, flux-linkage, torque and voltage quantities vs. rotor position for motor #1 for four different values of θ_c, going from θ_c^ψ to θ_c^i (left to right), at equal values of speed and total torque.

voltage). The r.m.s. current was checked not to exceed the maximum level at low speeds by iterative simulations. The results are shown in Figures 6.9 and 6.10.

The simulated speed data in Figure 6.9 and 6.10 deserve a degree of scepticism. The simulation programme calculates flux-linkage waveforms corresponding to a specified torque level τ_d, field angle θ_c, speed ω and supply voltage V_{DC}. It is assured manually

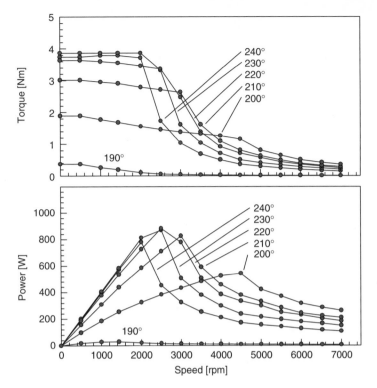

Fig. 6.9 Simulated torque and power vs. speed for motor #1 with constrained phase voltage ($V_{DC} = 100$ V) and r.m.s. current (10 A). $\theta_c = [190°, 200°, 210°, 220°, 230°, 240°]$. Note: $\theta_c^i \approx 235°$ and $\theta_c^\psi \approx 200°$.

that the r.m.s. current does not exceed the specified limit. Visual inspection of the phase voltage $v_{ph} = \omega \cdot d\psi/d\theta$ is necessary to determine whether $|v_{ph}| \leq V_{DC}$ for all rotor positions. The numerical differentiation of flux-linkage with respect to position causes problems similar to those in the torque computation. The voltage waveforms look rather noisy and admittedly the judgement of whether $|v_{ph}|$ is within the limits of the inverter voltage is subject to error. The simulation procedure is shown in Figure 6.11.

However, the simulation results for two example motors confirm that θ_c must be changed with the operating speed. At low speed maximum torque at maximum efficiency can be obtained using $\theta_c = \theta_c^i$ and the torque/speed range can be extended using $\theta_c = \theta_c^\psi$. A transition between the two extremes can assure a maximized torque/speed envelope. The switched reluctance motor drive will move from a current-limited operating mode, through a field-weakening transition, to a final operating mode in a way similar to the synchronous reluctance motor (Betz *et al.* 1993), and can theoretically achieve smooth torque (albeit at a low level) at infinite speed. The block diagram representation of the proposed torque controller is shown in Figure 6.12.

Other similarities include the effect of saturation. At $\theta = \theta_c^i$ the machine is more likely to be saturated than at $\theta = \theta_c^\psi$. Saturation moves θ_c^i closer to θ_c^ψ than for the unsaturated motor and therefore reduces the obtainable field-weakening range, which is consistent with the synchronous reluctance motor where an increased saliency ratio ξ also may enhance the drive's torque/speed capability (Soong and Miller, 1993).

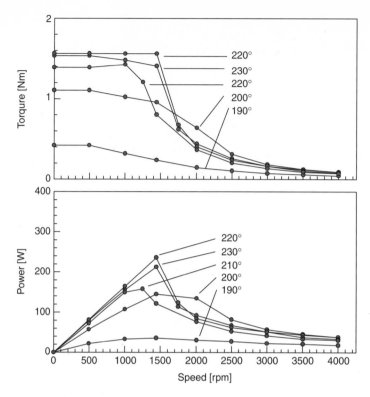

Fig. 6.10 Simulated torque and power vs. speed for motor #2 with constrained phase voltage ($V_{DC} = 100\,\text{V}$) and r.m.s. current (10 A). $\theta_c = [190°, 200°, 210°, 220°, 230°, 240°]$. Note: $\theta_c^i \approx 230°$ and $\theta_c^\psi \approx 195°$.

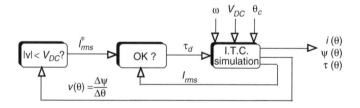

Fig. 6.11 Process for low ripple torque/speed simulations.

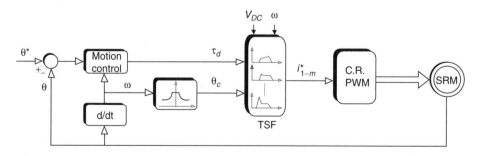

Fig. 6.12 Block diagram of instantaneous torque controller architecture.

6.4   Experimental results

Three features of the proposed controller are tested by experiment: that reduced torque ripple is achievable, that the shaft torque varies linearly with reference torque and that field-weakening can extend the speed range.

6.4.1   Torque ripple assessment

Switched reluctance motor #1 was connected to a PM d.c. commutator machine through a torque transducer and heavy backlash-free couplings. Three tests were conducted, using (in turn) conventional current control, the low-loss scheme and the low-flux-linkage scheme. The switched reluctance motor was operated with a constant torque demand (open-loop) but with no closed-loop speed control, because the speed feedback would have tended to suppress the effects of the torque ripple. Thus the speed was only approximately constant. For conventional control a turn-on angle of 44° (mech.) and a turn-off angle of 59° (mech.) were used. The phase currents and the torque transducer output were recorded and are shown in Figure 6.13.

The torque transducer waveform is strongly affected by drive train oscillations, as is most clearly indicated by the large transient in Figure 6.13 (middle), though the stroke frequency shows in Figure 6.13 (top) that other frequencies are present. It is believed that the heavy couplings together with the thin shaft of the torque transducer have made the readings invalid. Though the transducer gave correct running average torque values, it was found later that the particular torque transducer under load showed shaft deflections far beyond the manufacturer's specifications. This method of estimating torque ripple was consequently abandoned.

Instead the switched reluctance motor was connected directly to the load machine and open-loop tests were repeated. This time the rotor speed was captured over two mechanical revolutions using the incremental encoder feedback. The results are shown in Figure 6.14. The stroke frequency is dominant for conventional control whereas the speed ripple is reduced significantly for low torque ripple operation.

The encoder speed was differentiated to find acceleration, but the quantization and double differentiation of the 12-bit encoder signal yields results too noisy to be used for an absolute torque measurement. However, it is useful for comparison between the different excitation schemes. The r.m.s. values of acceleration were calculated from the encoder signal, and their comparisons

$$\frac{\alpha_{r.m.s.}(\theta_c^i)}{\alpha_{r.m.s.}(c.c.)} = 0.184$$

$$\frac{\alpha_{r.m.s.}(\theta_c^\psi)}{\alpha_{r.m.s.}(c.c.)} = 0.236 \tag{6.21}$$

indicate reductions in r.m.s. torque ripple of 5 and 4, respectively, when going from conventional current control (c.c.) to the proposed torque control scheme. Wallace and Taylor (1992) reported similar figures of improvement. Finally, the spatial harmonic spectra of calculated acceleration are shown in Figure 6.15. The stroke frequency (24 per mechanical revolution) and particularly its 2nd harmonic (i.e. 48 per mechanical

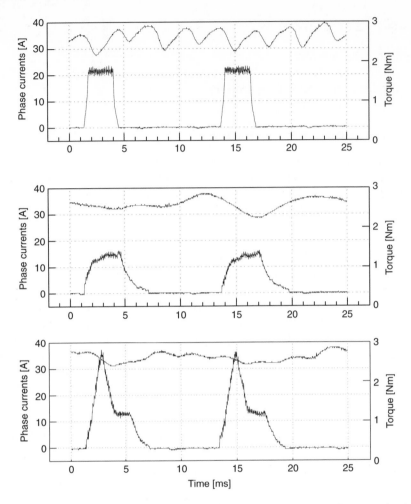

Fig. 6.13 Current and torque transducer output vs. time. Open-loop measurements for conventional control (top) and for low ripple torque control with θ_c^i (middle) and with θ_c^ψ (bottom).

revolution) are strongly reduced with the proposed torque control scheme:

$$\frac{\left|\alpha_{24}(\theta_c^i)\right|}{\left|\alpha_{24}(\text{c.c.})\right|} = 0.0096$$

$$\frac{\left|\alpha_{24}(\theta_c^\psi)\right|}{\left|\alpha_{24}(\text{c.c.})\right|} = 0.0521$$

$$\frac{\left|\alpha_{48}(\theta_c^i)\right|}{\left|\alpha_{48}(\text{c.c.})\right|} = 0.0138$$

$$\frac{\left|\alpha_{48}(\theta_c^\psi)\right|}{\left|\alpha_{48}(\text{c.c.})\right|} = 0.0329. \tag{6.22}$$

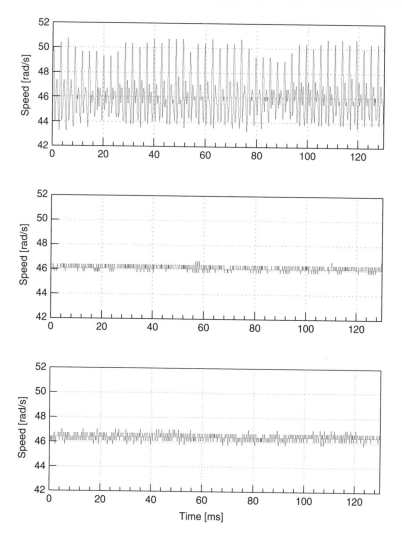

Fig. 6.14 Encoder speed vs. time. Open-loop measurements for conventional control (top) and for low ripple torque control with θ_c^i (middle) and with θ_c^ψ (bottom).

6.4.2   Alternative method for torque ripple assessment

Because of the difficulties of experimental determination of absolute torque ripple, a third approach was adopted (Kjaer, 1997b), and this new method is briefly described below. The conventional interpretation of torque ripple is to show how the servo motor's shaft torque τ_{shaft} varies vs. rotor position when subjected to a constant torque demand τ_{ref}. Instead, it is suggested depicting how the servo motor torque demand varies vs. rotor position when subject to a constant shaft (load) torque. This torque ripple measurement is performed at standstill with the test machine connected to a PM d.c. commutator motor. Controlling the armature current in the load motor the load torque can be accurately known.

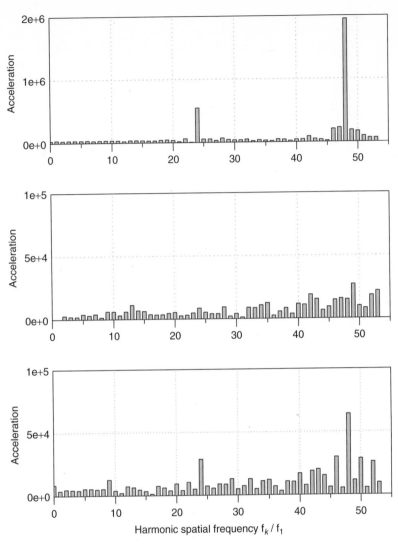

Fig. 6.15 Spatial frequency spectra of computed acceleration amplitudes. Open-loop measurements for conventional control (top) and for low ripple torque control with θ_c^i (middle) and with θ_c^ψ (bottom). Note that vertical scales differ.

   Applying closed-loop position control to the servo motor it may be stopped at any position. The motor is brought to steady-state (zero speed), while the user has access to the values of the servo motor torque reference, the rotor position and the torque of the load machine. Repeating the process for a range of rotor positions (ideally over one mechanical revolution), the servo drive torque reference variation vs. rotor position may be measured. A drawback is that the total load torque seen at the test motor shaft will include any coulomb friction present. Hence, care should be taken to minimize this. The number of commutator segments in the load machine should be large to avoid significant variation of the load torque. (Should a d.c. machine prove inadequate, an iron-powder hysteresis brake can be used instead (Holtz and Springob, 1996).) It could be argued that monitoring $\tau_{ref}(\theta)$ under running conditions with closed-loop speed control could be employed instead of the proposed standstill method. However, this would require a very high bandwidth speed control loop, in order that the instantaneous speed can be regarded as constant. Also, a truly constant load torque is required.

The proposed technique gives the variation in reference torque vs. position. The reference torque ripple factor may be calculated as:

$$K_{ripple,ref}(\theta) = \frac{\tau_{ref}(\theta) - \tau_{ref,avg}}{\tau_{ref,avg}}. \tag{6.23}$$

The servo drive was placed in closed-loop position control, and the proposed technique was applied over two electrical cycles in steps of $1°$ (mech.). Both the low-loss and the low-flux waveforms were tried (calculated for 3000 rpm operation). The results are shown in Figure 6.16.

Two numbers are of interest when assessing torque ripple: the absolute peak ripple measured from peak torque to average torque, and the mean absolute ripple. For both ripple calculations, these numbers are shown in Table 6.1.

Measurements with both TSF waveforms show the same difference between the average reference torque and the (constant) load torque (see below for linearity tests). The spatial periodicity of $15°$ corresponds to the number of pole-passings per revolution (the stroke harmonic frequency is 24), which gives some confidence in the experimental readings. There is some difference between the first $15°$ and the last $15°$ in both cases.

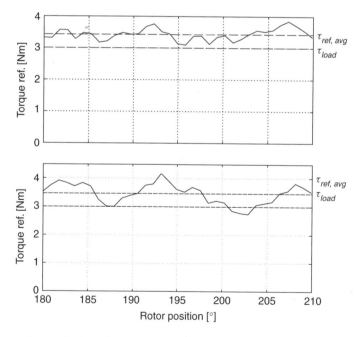

Fig. 6.16 'Inverse' torque ripple results. Measured variation of torque reference vs. position for constant load torque with motor #1 in closed-loop position control, using torque sharing functions $\theta_c = \theta_c^i$ (upper) and $\theta_c = \theta_c^\psi$ (lower).

Table 6.1 Torque ripple figures for reference torque measured by alternative method on motor #1

	$K_{ripple,ref}$	
	mean	max
$\theta_c = \theta_c^i$	4.27%	12.45%
$\theta_c = \theta_c^\psi$	8.68%	20.85%

This is believed to be due to either position dependent load torque (nonconstant d.c. motor torque constant or variable coulomb friction) or electromagnetic differences between phases. The results might have been of higher quality had an iron-powder brake been used for this test instead of the d.c.-machine. The actual source of these differences was not determined. Also worth noting is that the low-flux waveform shows larger torque variation than the low-loss waveform, possibly because operation at θ_c^ψ implies a lower torque-per-ampere with a higher sensitivity to modelling errors.

The ripple on the pulse-width modulated current waveform at standstill is at least $V_{DC}/(2 \cdot l(\theta) \cdot f_{sw})$, where $l(\theta)$ is the incremental inductance at the particular position. With a 100 V controller voltage and 19.2 kHz switching frequency, the current ripple with an inductance of (for example) 3 mH becomes approximately 1 A peak-peak. Hence, the phase currents can be regarded as specified by the torque controller and the remaining ripple may be due to poor torque calculation by the control law.

6.4.3   Torque linearity

The average shaft torque was measured with the torque transducer for different values of reference torque, running at 2000 rpm. Here, the offset in measured torque vs. reference torque is due to coulomb friction. A best first-order fit gives an offset of 0.57 Nm and a gradient of 1.0.

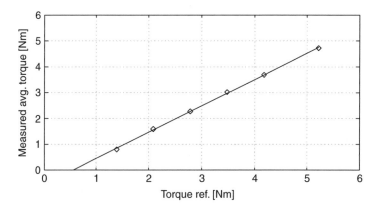

Fig. 6.17 Measured open-loop torque linearity for motor #1.

6.4.4 Measured torque/speed capability

With the torque ripple properties established the torque/speed capability was investigated experimentally. For the two extreme field-angles θ_c^i and θ_c^ψ the obtainable torque/speed capability is shown in Figure 6.18. Clearly, the predicted extension of the torque/speed capability is verified. Beyond the speeds recorded, the waveforms deteriorate (due to lack of voltage) and 'smooth' torque can no longer be assured[2]. As the maximum speed of the load motor was 3500 rpm, the switched reluctance motor voltage was reduced from its nominal 270 V to 100 V. The constant power speed range (CPSR) depends on where base-speed[3] is defined. From Figure 6.18 it appears to be around 1000 rpm giving a measured CPSR of over 3:1.

In the experiments, the TSF waveforms were calculated for maximum speed only (3000 rpm) for ease of implementation. This means that full utilization is not made of the controller voltage at lower speeds, but the effect of not tailoring the waveforms to more than one speed is not significant. At low speed $\theta_c = \theta_c^i$, and the current is low during the magnetization and demagnetization periods, but there will be a slight

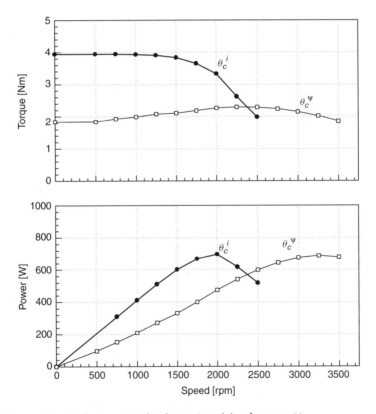

Fig. 6.18 Measured low ripple torque/speed and power/speed data for motor #1.

---

[2] This phenomenon corresponds to over-modulation in PWM operated a.c. drives.
[3] Base speed is the maximum speed at which rated torque can be obtained.

penalty in copper losses. For higher speeds, $\theta_c = \theta_c^{\psi}$ and the currents are higher, but so is the required voltage.

To compare the proposed torque control with conventional control, two sets of simulations were conducted, shown in Figure 6.19. At low speed similar torques can be obtained with the two schemes for the same r.m.s. current, but at high speed the requirement for smooth torque is not as efficient as, for example, single-pulse operation (high ripple). Essentially, low ripple with low flux requires the current to be boosted at angles where the torque-per-ampere ratio is very low. If low ripple is only a requirement for low speeds, then a change of control strategies should be considered.

6.4.5 Some comments on the torque control scheme

If mutual coupling is included in the electromagnetic model and the calculation of torque sharing waveforms, the commutation angles θ_c could still be calculated according to the criteria in (6.5) and (6.10), and the same analysis as shown here applies. If electromagnetic differences between phases or rotor eccentricity are

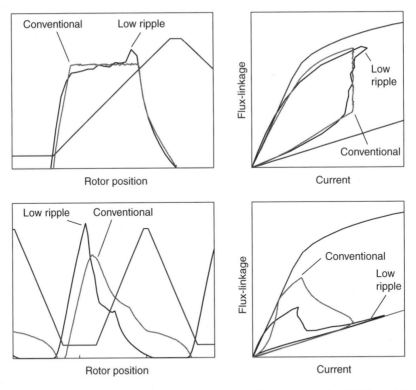

Fig. 6.19 Comparison between conventional control and low ripple torque control at 10 A (r.m.s.) simulated for motor #1. Running waveforms and (i, ψ)-loops. Top, low ripple: 1000 rpm, $\theta_c = \theta_c^i$, $\tau_d = 3.92$ Nm. Top, conventional: 1000 rpm, $\theta_{on} = 225°$, $\theta_{off} = 320°$, $i_{ref} = 20$ A, $\tau_{avg} = 4.02$ Nm. Bottom, low ripple: 6000 rpm, $\theta_c = \theta_c^{\psi}$, $\tau_d = 1.09$ Nm. Bottom, conventional: 6000 rpm, $\theta_{on} = 140°$, $\theta_{off} = 315°$, $i_{ref} = 20$ A, $\tau_{avg} = 2.32$ Nm.

significant the torque control scheme will prove inaccurate. As discussed elsewhere, the electromagnetic properties measured off-line at the terminals are used for off-line torque computation. A powerful DSP could measure magnetization characteristics *on-line* (monitoring current, rotor position and flux-linkage), and after an initial motor identification, calculate the required torque-sharing waveforms. This is a cumbersome task, but it would replace the need for pre-programming of the switched reluctance motor controller.

6.5   Application of ITC in servo-system

The remainder of this chapter examines the testing and assessment of high-bandwidth switched reluctance motor torque control for actuator and motion control applications. Of interest is whether the torque control is sufficiently fast and accurate that it does not, in itself, impose any significant limits on the performance of the overall system. The quality of torque control should permit the design of conventional velocity and position control loops in a transparent fashion. The objectives of this chapter are (i) to demonstrate that the switched reluctance motor can be used in actuator applications and (ii) that its nonlinear torque production and nonlinear torque controller do not impose significant performance restrictions. The following analysis also aims to determine a general test procedure to assess the switched reluctance motor's suitability for use in servo-systems, and investigates how to overcome the nonlinearities while making use of its advantages, that is:

(a) How should the switched reluctance motor be employed in high-bandwidth motion control applications? What system architecture should be used?
(b) What factors limit the potential performance in these applications? How can the actual performance be maximized?

As was discussed in the previous section, the problem of controlling the motor torque translates into that of controlling the phase currents to track their reference values. Figure 6.20 shows how a torque demand τ_{ref} is translated via torque-sharing functions into four reference phase currents. The closed-loop current-regulated PWM power converter, and the motor's nonlinear relationship of $i(\psi, \theta)$ and $\tau(i, \theta)$ are shown for a single phase, and the torque contributions from the four phases are added. Having established the block diagram from reference torque to motor torque, the following criteria were formulated in order to assess the quality of the torque control scheme:

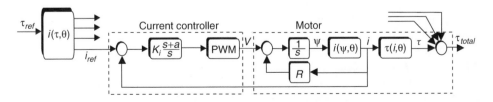

Fig. 6.20 Block diagram of torque controlled switched reluctance motor (only one of four phases shown).

(i) The torque control should not, in itself, impose significant limits on the performance of the overall system.

(ii) The torque control should be 'transparent' in the design of the velocity and position control loops. This means that it linearizes the relationship between reference torque and actual torque.

(iii) High-performance motion control of the load object should be achievable, in particular:

- variable speed control in four quadrants;
- low speed motion, including zero speed, with load torque present;
- high-bandwidth 'linear' response for small displacements;
- profiled motion.

In the following sections it is examined how to test the system against these criteria in order to provide 'proof-of-concept' of the proposed torque control scheme.

6.5.1 Inherent limits to performance imposed by torque control

With reference to criteria (i) and (ii) of the previous section, it can be claimed that the closed-loop torque control system of Figure 6.20 behaves essentially linearly and can be modelled in transfer function terms by the formula:

$$\tau(s) = k_\tau \cdot e^{-sT} \cdot \tau_{ref}(s) \tag{6.24}$$

This being so, the inner torque-control system will be transparent to the designer of position and velocity control loops if k_τ is close to 1, and T is sufficiently small, where k_τ is the ratio between output torque and reference torque, and T is the torque control time delay. k_τ and T are not available in analytical form and T in particular varies with the operating point, due to the nonlinearities of the switched reluctance machine. Instead, an assessment of the boundaries of both T and k_τ is attempted here. T is highly dependent on the time of response of the current regulating loops. The nonlinear electromagnetics of the switched reluctance motor make the electrical time constant of a phase winding vary both with rotor position and current level. A distinction is made between two methods to assess the torque (or current) control bandwidth. A linear small-signal analysis is made for small variations in phase current. For large-signal variations of torque the analysis is more appropriately done in terms of flux-linkage.

It should be made clear that (6.24) is intended solely to provide a starting point for assessing the impact of the torque control system on the design of the velocity and position loops. If k_τ and T are sufficiently close to their ideal values of 1 and 0 respectively, then these outer loops may be designed without particular regard to the details of the torque control. It is not suggested that (6.24) should be used beyond this context, e.g. as the basis of a simulation model.

Maximum time delay
A given torque demand may, at any rotor position, be translated into a corresponding flux-linkage value. The maximum time T_{max} it would take to bring the torque of a motor phase from zero to its maximum can be regarded (worst case) as the time

required to bring the flux-linkage from zero to its maximum value. An upper bound on this value may be read from the aligned flux-linkage at maximum phase current. For motor #1, $\psi_{max} = 0.15$ Vs. An estimate of T_{max} can then be expressed as the time it takes to bring the flux-linkage in a phase from zero to ψ_{max} or vice versa (neglecting the resistive voltage drop):

$$T_{max} = \frac{\psi_{max}}{V_{DC}} \tag{6.25}$$

where V_{DC} is the power electronic controller's supply voltage. T_{max} is the worst-case delay associated with large-signal dynamics, and is independent of the operating point. For the motor used in this work we get: $T_{max} = 0.15$ Vs$/150$ V $= 1$ ms. This number is very conservative, but furthermore, it would require a dead-beat flux-linkage controller to change the torque from zero to its maximum value in T_{max}. (The torque control developed in the previous section can minimize the time delay for a given torque by keeping ψ_{max} small, i.e. operating with $\theta_c = \theta_c^\psi$.)

Current control bandwidth – linear analysis

As the motor torque is directly related to the phase currents, the small-signal bandwidth of the closed-loop current regulating control will represent an upper boundary for T. The phase inductance varies with rotor position but has its maximum at aligned position θ_a. A worst-case linear analysis may then be to consider the phase winding as an LR circuit, where L is the unsaturated aligned inductance. PI regulation is adequate for current control, and the regulator gains may be decided based on different criteria. For the motor considered we get: $T_{elec} = L/R \approx 0.009/0.3 = 0.03$ s. A second-order current loop with a gain cross-over frequency of $\omega_{gc} \approx 4500$ rad/s (natural frequency $\omega_n = 5000$ rad/s) and a damping ratio of $\xi = 0.4$ was designed. As the design is based on the largest possible time constant, the closed-loop system will most often exhibit better damping and faster response.

The actual time delay of the torque control scheme will therefore be somewhere between T_{max} and $1/\omega_{gc}$. This time delay may be neglected in the design of the velocity and position loops provided the gain cross-over frequencies of the latter are not too high. A reasonable rule of thumb might be that the velocity loop gain cross-over frequency ω_{gc} should not exceed one-tenth of the reciprocal of the time delay. For motor #1 this gives $\omega_{gc} \leq 100$–300 rad/s.

Torque linearity and smoothness

The objectives of the torque control are to make the torque controlled switched reluctance motor's behaviour approach the one shown in (6.24). It should therefore be assessed how much the motor torque varies with rotor position (torque ripple) and whether k_τ varies with the torque demand (linearity). In Section 6.4 the torque ripple was measured to a maximum of 11% (see Figure 6.16). Figure 6.17 showed the measured average motor torque vs. reference torque, where the switched reluctance motor is operated in open loop at a constant speed of 2000 rpm. The linearity is clear and in this case k_τ is close to 1.0. The measured average torque equals the motor torque less friction (the d.c. motor was used as load), the latter causing the offset at low levels of reference torque.

With the boundaries of k_τ and T established, claim (i) is demonstrated. Claim (ii) is proven as far as the linearity of the torque controller goes. Essentially, velocity (and position) loops can now be designed as if the transfer function from reference torque to shaft torque was unity, provided the bandwidth of the velocity control loop does not exceed the bounds of the electromagnetic time constants. The actual implementation of outer loops will reveal how transparent the torque loop is to the user.

6.5.2   Dynamic test-rig

To provide experimental proof-of-concept, a test-rig was designed and built to emulate an actuator application in a laboratory setting. The test-rig, which is shown in Figures 6.21 and 6.22, was designed with both variable speed and position control in mind. The philosophy behind the design of the test-rig, the velocity and position loops and the test programme was to push the motor close to its limits by fully utilizing the available torque. In this case the maximum short-term motor torque, τ_{max}, has been limited to 8.5 Nm.

For variable-speed testing the switched reluctance motor was connected to a controlled d.c. commutator load machine using large inertia, backlash-free couplings. There is also the option of inserting a torque transducer between the two motors. As discussed in Section 6.4, the torque transducer was abandoned for dynamic testing where the instantaneous torque is of interest, and only used for measurements

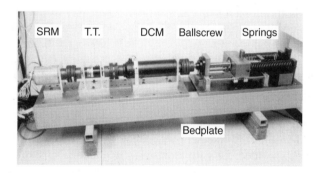

Fig. 6.21 Photograph of test-rig.

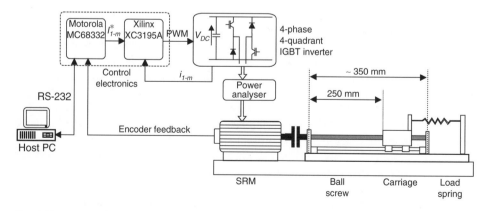

Fig. 6.22 Diagram of test-rig.

of average torque in steady-state operation. Figure 6.16, which is based on static measurements, shows that the torque reference signal is sufficiently close to the actual instantaneous torque that it is justifiable and useful to show the former in the experimental results.

For position control, the switched reluctance motor was connected via couplings to a ballscrew, on which a translational movement of a heavy carriage is possible. The ballscrew pitch p was 10 mm/turn, and the maximum possible travel S_{max} was 250 mm. To provide holding torque, tension springs of different sizes were used. The maximum torque derived via the ballscrew by the spring force results in a holding torque τ_{hold} of 4.0 Nm (maximum thermal limit) at a displacement $S = 206$ mm. This leaves for acceleration and slewing (in trapezoidal moves) a dynamic torque of $\tau_{dyn} = \tau_{max} - \tau_{hold} = 8.5 - 4.0 = 4.5$ Nm.

6.5.3   Design of velocity control loop

In this analysis the torque controlled switched reluctance motor drives the ballscrew and its carriage, and is enclosed by a velocity loop cascaded by a position loop. The block diagram of the system is shown in Figure 6.23. For small-signal analysis the spring load force has been regarded as constant.

For operation of the torque controlled switched reluctance motor in closed-loop speed control, PI speed regulation is adequate (with pure integral action the damping is poor for higher bandwidths). The system then resembles a standard second-order system with a zero:

$$\frac{\omega(s)}{\omega_{ref}(s)} = \frac{\frac{K_\omega}{J}(s+a)}{s^2 + \left(\frac{B + K_\omega}{J}\right)s + \frac{K_\omega a}{J}} = \left(\frac{\omega_n^2}{a}\right) \cdot \frac{(s+a)}{s^2 + 2\omega_n \zeta s + \omega_n^2}. \tag{6.26}$$

The PI-regulator parameters K_ω and a can be chosen as suggested by Dorf (1980), i.e. by specifying the damping ratio ζ, the natural frequency ω_n and the maximum overshoot M_r. One way to test the switched reluctance motor to its safe limits is to monitor the closed-loop speed control's transient response to step commands. If the torque demand τ_{ref} never must exceed τ_{max}, and the damping ratio and the maximum overshoot are kept constant, then the natural frequency ω_n may be plotted against the maximum magnitude of input steps $\delta\omega$ that just saturate the torque reference. Recalling that a holding torque may be present, the torque limit becomes $\tau_{dyn} = 4.5$ Nm. The results are plotted in Figure 6.24 for the motor used here with a damping ratio $\zeta = 0.70$ and maximum overshoot $M_r = 20\%$.

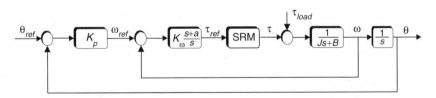

Fig. 6.23 Block diagram of position and speed control with idealized switched reluctance motor control.

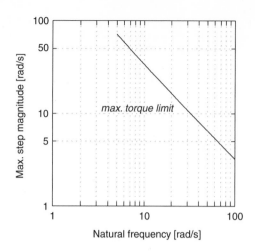

Fig. 6.24 Maximum small-signal step input amplitude vs. natural frequency of closed-loop speed controlled system for fixed damping and peak torque.

Regulators with these properties for damping and overshoot have been tested experimentally, with the encoder feedback used for speed detection. The finite resolution of the encoder is responsible for some variation in estimated speed even at steady-state, and it was found that there is a trade-off between dynamic performance and the noise in steady-state operation. That is, the greater the value of ω_n, the smaller the step size $\delta\omega$ that will just saturate the torque, and as the speed of response is increased, the signal-to-noise ratio of the measured response gets worse. The minimum amplitude of step-inputs was chosen to be limited to 10 times the steady-state variation in detected speed. This, in turn, limits the maximum natural frequency to 50 rad/s. Higher values are achievable, but with worse signal-to-noise ratios.

Measured examples of the closed-loop transient response of the speed controlled system are shown in Figure 6.25, for speed controllers yielding system natural frequencies of 20, 50 and 100 rad/s, respectively. With the springs disconnected the motor moves the load at 44 rad/s where the steady-state load torque essentially is constant at around 2 Nm. The system is then subjected to steps in desired velocity of the maximum permissible magnitude. Clearly, during transients the reference torque signal does not exceed $\tau_{hold} + 4.5$ Nm, but pushes the small-signal operation of the torque control just to its limits. Note that the transient responses in all three cases are very similar and closely resemble traditional, linear analysis predictions.

For this particular closed-loop system, it is evident that the switched reluctance motor torque control itself does not impinge any restrictions on performance. Rather, the velocity control bandwidth is limited by the moment of inertia and by variation in the encoder feedback signal. We then get: $\omega_n = 50$ rad/s, $K_\omega = 0.6$ and $a = 36$ rad/s as the fastest achievable closed-loop velocity response.

The derating of the S/N ratio with bandwidth is essentially what determines the fastest velocity response. Had the position feedback been of higher quality the torque loop ripple and current loop bandwidth *could* have compromised the speed loop. However, it is rare in servo drives that the mechanical load is such that the inner loops limit the dynamic performance.

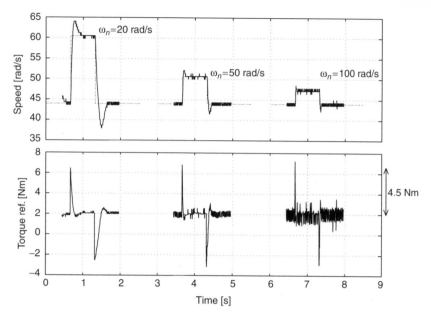

Fig. 6.25 Measured closed-loop response to small-signal steps in reference speed, with torque reference remaining unsaturated. Speed vs. time (top) and reference torque vs. time (bottom).

6.5.4   Design of position control loop

With the speed control loop designed, the small-signal position control can be analysed in a similar manner. The previous section defined $\delta\omega$ as the maximum magnitude of a step in reference speed that could be imposed without the torque reference saturating. If the reference position receives step commands of a magnitude δx, and the torque reference is to remain unsaturated, then the product $K_p \cdot \delta x$ must not exceed $\delta\omega$. Various combinations of K_p and δx were tried, and samples of the small-signal behaviour are shown in Figure 6.26. The holding torque in these experiments is about 3.9 Nm, and as shown the torque reference just saturates. For a position step of 2.50 mm (90° mech. or 1024 encoder pulses) a response time of 0.75 s is achieved, whereas for a position step of 0.31 mm (11.25° mech. or 128 encoder pulses) the response time is less than 0.1 s. Again, these exponential responses are close to the predictions of traditional linear theory.

Finally, to confirm the design approach used here, small steps in position reference were applied at different carriage positions, i.e. with different holding torques. The controller is expected to deliver similar transient responses anywhere within a large range of position, from $x = 0$ to $x = 218$ mm. The measured responses are shown in Figure 6.27. The closed-loop position response is clearly independent of the starting position (and static load), except at very large spring tension (where the load torque is higher than expected due to larger friction around zero speed) and where the torque signal saturates.

This section concludes that linear analysis and operation of small-signal motion control loops can be performed satisfactorily with the switched reluctance motor and

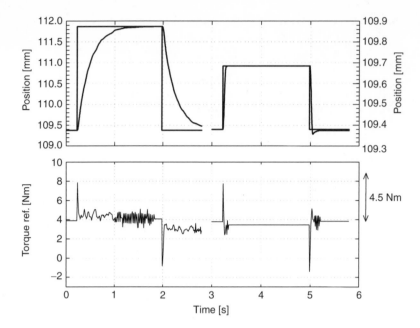

Fig. 6.26 Measured closed-loop position control response to small-signal steps in reference position, with unsaturated torque reference. Position vs. time (top); reference torque vs. time (bottom). Left: $\delta x = 2.50$ mm, right: $\delta x = 0.31$ mm.

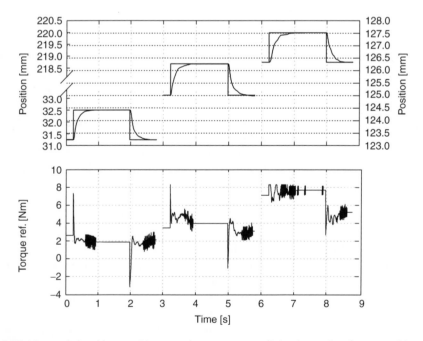

Fig. 6.27 Measured closed-loop position control response to small-signal steps in reference position, with unsaturated torque reference. Position vs. time (top); reference torque vs. time (bottom). $\delta x = 1.25$ mm, at starting positions $x_0 = 31.25$ mm, 125.00 mm and 218.75 mm.

its torque controller modelled as in (6.24). The electrical time constant normally being much smaller than the mechanical time constant permits the switched reluctance motor torque control of Section 6.3 to be assumed ideal.

6.5.5 Profiled motion control operation

Whereas the previous section focused on linear small-signal behaviour of the switched reluctance motor-based actuator, this section will demonstrate that the switched reluctance motor can be operated in a typical actuator duty.

Figure 6.28 shows a measured large-signal translational movement from $x = 31.25$ mm to $x = 218.75$ mm and return with the tension springs mounted. The acceleration limit is set by the controller to 318 rad/s^2 and the slewing speed to 87 rad/s. Higher values of $\omega_{slew,max}$ are limited by the maximum length of travel, but the acceleration is close to the predicted maximum. Clearly, trapezoidal operation is possible as well as good variable-speed operation in all four quadrants.

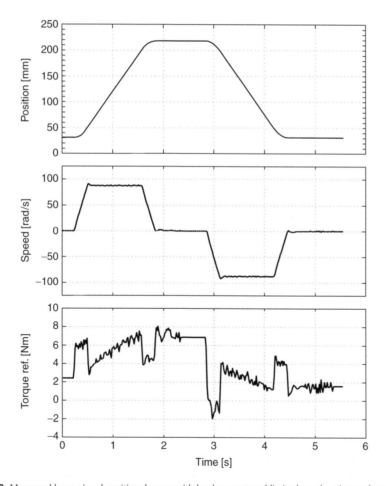

Fig. 6.28 Measured large-signal positional move with load present and limited acceleration and speed.

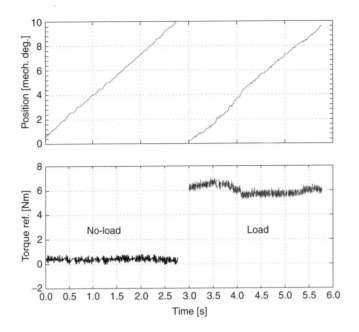

Fig. 6.29 Measured drive behaviour at 0.5 rpm.

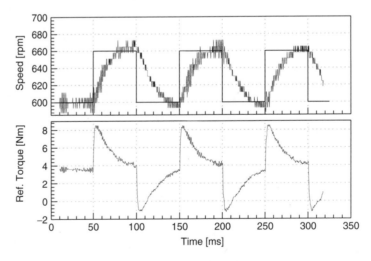

Fig. 6.30 Closed-loop speed control with d.c. motor as load. Transient response to 10 Hz square-wave speed reference.

Operation at 0.5 rpm in closed-loop speed control with and without load torque applied is shown in Figure 6.29. The speed is not particularly smooth at these frequencies due to the limited resolution in the encoder feedback signal, and possibly position dependent friction torque. Any motor torque ripple there might be is unimportant as the closed-loop speed controller compensates it immediately.

6.5.6   Variable-speed measurements

With the d.c. motor as load (a somewhat higher moment of inertia) a closed-loop speed
controller (different from the one used with the ballscrew) was tested. The transient
response to a 10 Hz square-wave on reference speed is shown in Figure 6.30. The
torque signal does not saturate and approximating the linear response to a first-order
system indicates a 15–20 Hz small-signal bandwidth in this case (Dorf, 1980).

   Finally, a load torque disturbance rejection test was conducted. With the d.c. machine
in closed-loop current control, the load torque was stepped up and down, while the
switched reluctance motor was in closed-loop position control. Recovery of position in
a few tenths of a second with less than 10°(mech.) excursion are shown in Figure 6.31.

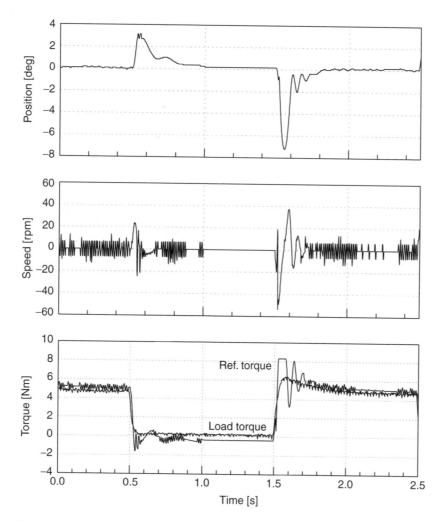

Fig. 6.31 Load disturbance rejection measurements: rotor position (top), speed (middle) and reference and
load torques (bottom) vs. time.

6.6   Discussion and conclusion

In Sections 6.1–6.4, a new theory for high-performance torque control has been presented. Low ripple operation with profiled currents must replace conventional commutation for high-bandwidth torque control. The lack of inherent torque sharing between phases requires the application of secondary criteria in order to determine the torque waveform for a single phase. Low-loss and low-flux operation define the extremes of the possible operating range, and to move between the two a variable field-angle has been introduced, allowing seamless transition from high-efficiency into field-weakening operation. Maximum inverter utilization is assured by tailoring the waveforms to the inverter voltage and operating speed.

A reduction of torque ripple by a factor of 5 compared to conventional commutation control is measured for the new torque control scheme, using a new method for torque ripple assessment, as other methods had to be abandoned. Experiments confirm the extended smooth torque operating range with field-weakening. Section 6.5 has analysed the implications of employing switched reluctance machines in actuator applications and how to assess the quality of the torque control. It also reports on a series of dynamic tests that can be used to push the motor and its torque control to its safe limits and to assess the achievable performance.

It has been shown that the torque control is sufficiently fast and accurate that it does not impose any significant inherent limits on the performance of the overall system. It has been demonstrated that the torque control itself does not impinge on the design process of motion control loops, which may be designed in the usual manner (claim (i) in section 6.5). The switched reluctance motor and its torque controller were transparent to the design of velocity and position loops. Upper boundaries for the bandwidth of these outer loops are derived, which may serve as general guidelines in the design, and high-bandwidth small-signal operation is demonstrated (claim (ii)). Finally, large-signal profiled motion as well as low-speed operation is reported (claim (iii)). It may be concluded that the switched reluctance motor and a torque controller of this type are amenable to motion control applications traditionally served by conventional servo motors.

7

Sensorless control

Gabriel Gallegos-López

The elimination of the mechanical position sensor by an electronic method is known as *sensorless control*. Many interesting sensorless methods for the switched reluctance motor have been proposed by researchers over the years. Despite advancements in sensorless control strategy, none of the present schemes has been fully able to replace the mechanical sensor without putting some limitations in the drive. This chapter reviews the state of the art of existing rotor position estimation methods for switched reluctance motors. It is important to note that while most of the sensorless methods have been published for motor application, they may also be applied for generators. In the following sections, the sensorless methods are classified and their basic principles are discussed. Furthermore, their main advantages and disadvantages are also given.

7.1 Introduction

In recent years the switched reluctance motor has received considerable attention for variable-speed drive applications. Its simple construction makes it an interesting alternative to compete with permanent magnet brushless d.c. motor and induction motor drives. To obtain optimum performance, it is necessary to commutate the excitation current from phase to phase synchronously with the rotor position. This has usually necessitated the use of a mechanical position transducer attached to the motor shaft. Hall sensors, optical sensors together with slotted disc, encoders or resolvers attached to the shaft are normally used to supply the rotor position signal. However, recently there has been enormous interest in eliminating the mechanical rotor sensors mainly for two reasons:

1. *Reduction of cost*: The mechanical sensor may be a significant part of the overall system cost in cost-sensitive applications for fractional-kW drives. The size of the mechanical sensor and the required number of leads also play an important role in this kind of application, where the resolution required in the rotor position is usually low.
2. *Operation in a harsh environment*: In applications where the motor is operated in extreme environmental conditions, such as high pressure, temperature, humidity and extremely high speeds, the use of a mechanical sensor may lead to reliability problems. Usually, high resolution in rotor position and robustness are necessary for this kind of application.

Other factors that motivate the elimination of the mechanical sensors are the reduction in the overall physical envelope and the weight of the motor drive. Furthermore, the requirement for mechanical mounting of the rotor position transducer complicates the drive design, increases the requirement for maintenance, and is a possible source of failure.

7.2 Classification of sensorless methods

The fundamental principle of most sensorless methods is that the rotor position information can be obtained from the stator circuit measurements (current and voltage at the motor terminals) or their derived parameters (phase inductance or others). In other words, the position information is obtained basically from the magnetic characteristics of the machine itself. The magnetic characteristic of the switched reluctance motor is nonlinear and is especially influenced by the local saturation of the stator and rotor poles.

The magnetic characteristic of a switched reluctance machine is usually represented by the magnetization curves (see Chapter 3).[1] A set of measured magnetization curves of a four-phase 8/6 machine is shown in Figure 7.1. It can be observed that the curve at the unaligned position θ_u is a straight line. In contrast, the curve for the aligned position θ_a shows a deflection around 5 A where the magnetic saturation effect starts. Notice that the intermediate positions present two deflections, the first one around 5 A which is mainly caused by the local magnetic saturation of the stator and rotor pole corners, and the second one at higher current level depending on the rotor position. The main cause of the second deflection is the magnetic saturation of the stator and rotor

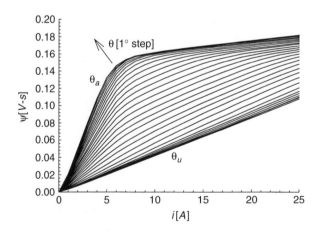

Fig. 7.1 Measured magnetization curves of a 5 kW 8/6 four-phase switched reluctance motor. The magnetization curves were measured using the direct method explained in Chapter 9.

---

[1] Half an electrical cycle is enough to represent the whole electrical cycle for a regular motor, i.e. from θ_u (the unaligned position, minimum inductance) to θ_a (the aligned position, maximum inductance) because θ_u is in the middle of the electrical cycle. In contrast, the whole electrical cycle is needed for an irregular motor because θ_u is not exactly in the middle of the electrical cycle (Miller, 1993). For magnetization curve measurement see Cossar and Miller (1992).

yoke. It is clear from the magnetization curves that the relationship between current, flux-linkage and rotor position is unique but strongly nonlinear which complicates the rotor position estimation.

From the phase voltage equation (7.1),[2] it can be observed that the incremental inductance l appears as a function of current and position, and therefore position estimation may be obtained from it. However, it is not an easy task due to the saturation effect.

$$v(i, \theta) = R \cdot i + \frac{d\psi(i, \theta)}{dt} \tag{7.1}$$

$$= R \cdot i + \left[\frac{\partial\psi(i, \theta)}{\partial i} \right]_{\theta=cst} \cdot \frac{di}{dt} + \left[\frac{\partial\psi(i, \theta)}{\partial \theta} \right]_{i=cst} \cdot \frac{d\theta}{dt}$$

$$= R \cdot i + l(i, \theta) \cdot \frac{di}{dt} + e.$$

The first term of equation (7.1) corresponds to the voltage due to the phase resistance R; the second term is the contribution of the inductive voltage, and the third term corresponds to the back-EMF e. The second and third terms vary strongly as a function of current level and position. Accurate measurements of the inductive voltage drop and the back-EMF are difficult when the motor is running, as both change significantly as a function of the motor operating point.

Figure 7.2 depicts the incremental inductance in two electrical cycles calculated from the set of magnetization curves of Figure 7.1. Note that l is not constant at any specific

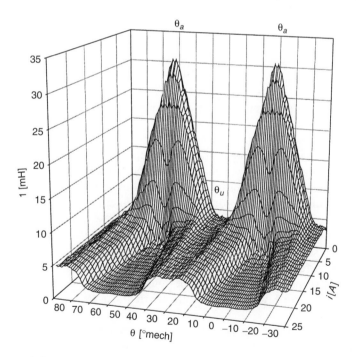

Fig. 7.2 Incremental phase inductance calculated from data in Figure 7.1.

---

[2] The mutually induced voltage is assumed to be zero.

position (except at θ_u) but it is decreased as a function of current, in particular at and near the aligned position. At high current levels it can even be smaller at θ_a than at θ_u. This results in an ambiguity for rotor position estimation if the machine is working in the saturation region, which is usually the case.

While most of the sensorless methods are based on the magnetic characteristic of the switched reluctance motor, there are many ways in which they may be classified. A straightforward classification method is to consider if the method is based on variables of the energized phases, variables of the unenergized phases or use of other variables. Using this criterion, the sensorless methods mainly fall into three major groups.

1. Open loop methods.

   - Dwell angle compensation.
   - Commutation angle compensation.

2. Energized phase methods.

   - Chopping waveform.
   - Regenerative current.
   - Flux-linkage.
   - State observers.
   - Irregularities in stator/rotor poles.
   - Current waveform.

3. Unenergized phase methods.

   - Active probing.
   - Modulated signal injection.
   - Regenerative current.
   - Mutually induced systems.

Before going into the detail of each sensorless method, a brief overview of each group is given.

The first group, also called stability torque control, does not provide a direct position indication. The motor works in open-loop as a stepper motor from a variable frequency oscillator in a traditional synchronous manner, and adjustment is made to the dwell angle or commutation angle by observing the variation in the d.c.-link current in order to improve stability in pull-out torque with maximum efficiency.

The second group uses the excitation current, voltage waveforms or derivatives (i.e. variables from the phase that is generating torque) to detect the rotor position indirectly. An early method senses di/dt, which is a function of the rotor position, when the motor is in hysteresis current regulation. Other methods, that belong to this group, make use of the regenerative phase current when the rotor crosses the aligned position. These methods mainly are for low speed where the current can be regulated (i.e. below the motor base speed).

Alternatively, there are methods in this group involving the detection of flux-linkage or phase inductance when the rotor passes through a particular threshold value which represents a specific rotor position. These methods may provide more resolution if the number of reference values is increased instead of having only one threshold value. An example of these methods is the method based on flux-linkage observer or current observer, which gives instantaneous rotor position estimation. Similarly, a complex lumped parameter reluctance network model of the motor has been proposed, which also estimates instantaneous rotor position. This group also includes state observer methods, which are based on a mathematical motor model that is run in real time. The methods that use flux-linkage suffer at low speed, where it is difficult to calculate it accurately because of the variation in the phase resistance and noise in the system. Conversely, methods that include a full model of the motor may be limited at high speed because of the large computational burden. In other words, the speed of the digital signal processor may limit the use of these sensorless methods for high-speed applications.

Other methods include irregularities in the stator poles, rotor poles or both. In this way, the inductance profile is modified so that when the phase is excited, it is easy to detect the irregularity in the phase current which indicates a specific rotor position. The last subgroup of this group uses the shape of the current waveform to estimate the position. A method has been proposed that identifies if the commutation angle is leading or lagging based on the shape of the current waveform. The commutation angle is compensated in order to have the desired current shape. However, this method does not estimate any specific rotor position. The methods *current gradient* and *voltage magnitude sensorless method* are based on the current and voltage waveform respectively. These methods identify one specific rotor position per stroke and they are simple to implement.

In the third group, different kinds of test signals are introduced or measured during the time when a phase is normally unenergized (i.e. the phases that are not generating torque), usually during the negative slope of the phase inductance when the machine is motoring. Note that when the machine is generating the positive slope of the phase inductance is used instead, because this is the region where the phase is not conducting.

The test signal needs to be of low amplitude for the following reasons (Ray & Al-Bahadly, 1993):

- to minimize negative torque production;
- to avoid saturation effects;
- to minimize back-EMF effects;
- to minimize the power rating of additional injection circuitry where this is necessary.

The principle of these methods is to detect the phase inductance or flux variation from the injected signal. The methods that belong to this group are: active probing, modulated signal injection, regenerative current and mutually induced systems. These methods suffer at high speed where there may not be enough time to inject the probing signal between the extinction and turn-on of the phase, because the phase current waveform is nonzero over the majority of the electrical cycle. Furthermore, the probing signal is sensitive to mutual interference from excitation currents in other phases.

The basic principle, main advantages and disadvantages, speed range,[3] dynamic capability, computational overheads, and resolution of the three major groups of sensorless methods are discussed as follows.

---

[3] The terms low speed, medium speed and high speed are defined as speed ranges from 0 to $\frac{1}{3}$ base speed, from $\frac{1}{3}$ base speed to base speed, and greater than base speed respectively.

7.3   Open-loop methods

In these methods the motor is running synchronously from a variable frequency oscillator. The basic principle is to improve the stability of open loop drive by adjusting the dwell angle or advancing the commutation angle based on the d.c.-link current.

When the machine is operated in open loop, the motor frequency and the dwell angle are controlled, but load torque fixes the torque angle.[4] Miller and Bass (1986) and Bass *et al.* (1986) show that the maximum pull-out torque is proportional to the dwell angle, i.e. at maximum dwell angle there is maximum pull-out torque. Also it is shown that the efficiency is inversely proportional to the dwell angle. In other words, at minimum dwell angle there is maximum efficiency. Furthermore the two maxima (pull-out torque and efficiency) are around the same torque angle. Hence, it was proposed that the dwell angle is adjusted to maintain a constant torque angle which maximizes the efficiency for different load torques. When there is a load transient the dwell angle is increased to compensate the load torque maintaining the torque angle.

The feedback signal is the average d.c. link current which is used to adjust the dwell angle, i.e. with an increase in d.c.-link current, the dwell angle must be increased in order to maintain the torque angle. To improve the stability still further, the frequency oscillator is also adjusted when there is a transient in load torque. Figure 7.3 shows the block diagram of the implementation. Oldenkamp (1995) made some improvements to this method by adding a controller circuit which allows changes in the speed and direction of rotation.

Vukosavic *et al.* (1990) proposed a similar method. The main difference is that the dwell angle is fixed and the commutation angle is advanced. The feedback signal is

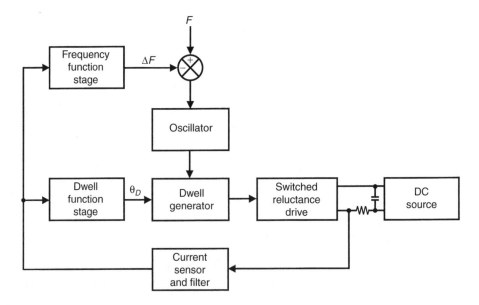

Fig. 7.3 Block diagram for stabilized control by dwell angle control.

---

[4] The torque angle is measured from the turn-off angle to the aligned position.

the average energy returned to the power supply, i.e. the current that flows through the power converter diodes.

The main advantages of these methods are: maximized efficiency with improved stability compared with the open-loop control, and low-cost implementation. The main disadvantage is that it is poor in dynamic performance because of the nature of the feedback signal, therefore it is not applicable for variable-speed drives. It appears that it is applicable to motors working at constant speed with approximately constant load. Zero speed is not possible and the method does not provide a direct rotor position signal since it assumes synchronism of the firing angles with the rotor position.

7.4 Energized phase methods

The methods that belong to this group are those which make use of variables from the phase that is generating torque. The first two major publications concerning indirect detection of rotor position were published by Acarnley *et al.* (1985) and Hill and Acarnley (1985) who proposed three methods. Two of them fall into this group and are explained later in this section. Another method in this group uses the regenerative current when the rotor passes the aligned position. Alternatively, methods that use the intrinsic magnetic characteristics of the machine belong to this group. Furthermore, methods based on lumped parameter network and state observers that make use of the energized phase for detecting rotor position are included in this group. On the other hand, this group also includes a method which introduces irregularities in the inductance profile in a specific rotor position, and others that use the shape of the current waveform or its derivative. All these methods are explained in detail as follows.

7.4.1 Chopping waveform

Acarnley *et al.* (1985) and Hill and Acarnley (1985) proposed this method in which the motor is controlled by hysteresis current regulation. The current is maintained approximately constant by a chopper in a hysteresis band $I \pm (\Delta i/2)$ as shown in Figure 7.4. The current swings around the required level at a rate dictated by the incremental inductance l. As the incremental inductance is rotor-position dependent, the instantaneous

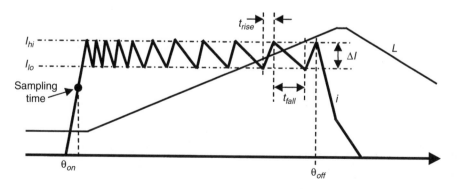

Fig. 7.4 Chopped current in active phase for waveform detection method.

rotor position may be detected indirectly from the chopping characteristics. Panda extended Acarnley's work and proposed the equations for the rise and fall time in the hysteresis band (Panda and Amaratunga, 1993a).

Rearranging equation (7.1) we have:

$$t_{rise} = \frac{l(i,\theta) \cdot \Delta i}{v - R \cdot i - \left[\frac{\partial \psi}{\partial \theta}\right]_{i=cst} \cdot \omega} \tag{7.2}$$

$$t_{fall} = \frac{l(i,\theta) \cdot \Delta i}{R \cdot i + \left[\frac{\partial \psi}{\partial \theta}\right]_{i=cst} \cdot \omega} \tag{7.3}$$

It can be observed from equations (7.2) and (7.3) that both t_{rise} and t_{fall} are functions of l and therefore the position can be detected assuming that the current swings in a small band relative to the mean current level. However, there are other variables which appear in these formulas, for instance the voltage drop across the phase resistance and the back-EMF which vary significantly as a function of speed and rotor position. The uncertainty in the phase resistance and back-EMF complicates rotor position detection using this principle. In practice $v \gg R \cdot i$ and $R \cdot i$ can be neglected. However, the back-EMF, which can be comparable to the voltage supply at high speeds, cannot be neglected. Furthermore the incremental inductance is current dependent which contributes to ambiguity and extra complexity of rotor position detection (Figure 7.2). The use of t_{rise} for detecting rotor position is preferable instead of t_{fall} (Panda and Amaratunga, 1993a). The reason is that t_{rise} depends on v unlike t_{fall}, and therefore the back-EMF and $R \cdot i$ can be neglected at low speed. However, this assumption will not apply at high speed and should be compensated somehow. A further study of this method was done by Panda and Amaratunga (1993a), Panda and Amaratunga (1993b) and Panda and Amaratunga (1991b) and the results showed the difficulty of estimating the position with this method. A comparison between the open-loop method and the chopping current waveform method is shown in Panda and Amaratunga (1991a) which suggests some modifications to improve the stability of the method.

Alternatively, Kalpathi (1998) proposed in a recent patent to use the ratio between the rise and fall times to estimate the incremental inductance and then the rotor position. This method makes use of hysteresis current control with soft chopping. The incremental inductance is expressed as,

$$l = \frac{v}{\Delta i} \cdot \frac{T_r \cdot T_f}{T_r + T_f} \tag{7.4}$$

where T_r and T_f are the rise and fall times respectively. As in previous methods, the phase is commutated when the time ratio reaches a predetermined threshold value.

The main advantage of these methods is that they can be implemented with simple electronics. The main application may be for drives <1 kW at low speed where current regulation is possible. The major problems of this method are: the back-EMF should be known which is difficult to measure while the motor is running, the fact that the incremental inductance also depends on the current amplitude introduces ambiguity at high current levels, and the assumption of current chopping limits these methods to low speeds. The resolution of rotor position estimation that can be achieved with this method depends on the number of current chops that occur in one energy conversion stroke. This means that the resolution is decreased in proportion to the speed and therefore, instantaneous rotor position estimation is very difficult.

Holling *et al*. (1997) and Holling *et al*. (1998) patented a sensorless method that also uses the rate of change of the phase current (di/dt), but they propose to apply PWM to the motor phases and monitor di/dt in each PWM period. di/dt is a function of the incremental inductance as it can be observed from equations (7.2) and (7.3). Hence position can be estimated. However, it is clear that this method also suffers from the ambiguity of the incremental inductance at high current levels and errors introduced by the back-EMF. Therefore, the position estimator has to be compensated as a function of speed and current level in order to obtain an accurate position estimation. Another disadvantage is that single pulse operation is not possible. However, the main advantage is that it allows PWM current control.

The second method proposed by Acarnley *et al*. (1985) is the use of the initial current gradient rather than the rise time in the chopping waveform. Obradovic (1988) extended this idea and proposed to sample the current at a predetermined time after the beginning of the commutation as shown in Figure 7.4. The current at that time is a function of rotor position governed by equation (7.2) (the drive still being controlled by hysteresis current control). The sampled current value is compared with a reference value representative of rotor position and the difference between them is used to control the hysteresis current reference in order to correct the desired rotor position during the next cycle. Feedback compensation is necessary to give acceptable stability. Using the current gradient at the beginning of commutation for position estimation has the advantage that the effect of the back-EMF and saturation is minimized because the sampling is done close to the unaligned position, where the phase inductance is linear. However, the method is still limited to low and medium speed, where current regulation is possible.

7.4.2 Regenerative current

This method, proposed by Reichard and Weber (1989), senses the regeneration current in the energized phase. The phase is de-energized in response to such current. In this method, the drive is controlled by current control with the special characteristic that when the current is above the current reference the phase is turned off for a constant period of time (i.e. no hysteresis control). The control is in soft chopping and the freewheeling current through the diode is observed. When the rotor is in the region where the inductance is increasing, normally the freewheeling current decays along a negative slope; however, when the rotor passes the aligned position the inductance starts to decrease, therefore the negative slope of the freewheeling current is interrupted and starts to increase along a positive slope. This change from negative to positive slope of the freewheeling current indicates when the rotor passed the aligned position. Figure 7.5 shows a typical current waveform used in this method.

The main disadvantage of this method is that the energized phase is normally turned off just after the aligned position, where considerable negative torque may be generated and this may result in poor efficiency of the drive. Other drawbacks are that the turn-off angle cannot be advanced, and the speed range is limited to speeds below the base

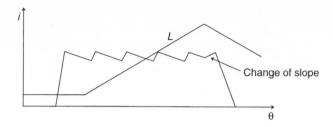

Fig. 7.5 Chopped current in active phase for regenerative current detection method.

speed where current regulation is possible. This method may be applicable to small motors at low constant speed operation. However, it should be noted that this method is not applicable at standstill.

The main advantage of this method could be the easy implementation.

7.4.3 Flux-linkage

Several methods have been proposed which make use of pre-stored values of either phase inductance or flux-linkage for detecting rotor position. For instance, Hedlund and Lundberg (1991) proposed to identify a particular inductance value per stroke which corresponds to a rotor position in the rising inductance region. The signal which identifies the specific rotor position is created from the current in and the voltage across the phase winding without the motor drive being affected directly by the measuring process. According to the method, a known position is reached when the inductance L of the phase winding in question reaches a predetermined inductance value L_{ref}. The condition $L > L_{ref}$ is calculated by comparing the integrated voltage in the phase winding to the product of the value of the actual phase current and the predetermined inductance L_{ref} as shown in equation (7.5).

$$\psi = \int (v_{ph} - R \cdot i) \, dt > L_{ref} \cdot i. \tag{7.5}$$

The block diagram of the method is shown in Figure 7.6. The method uses two extra compensation factors implemented in the microcontroller box (μP). The current-dependent factor corrects the errors caused by the fact that the phase inductance varies as a function of the phase current (this may be called saturation factor). The second

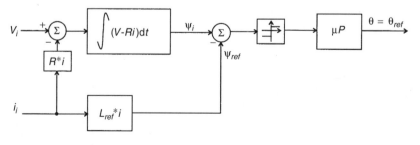

Fig. 7.6 Block diagram of Hedlund's method.

factor introduces a delay for phase commutation as a function of speed (this may be called the back-EMF factor). It is important to note that this method uses only one inductance value as a reference. However, the method needs two compensation factors that, in fact, are two vectors which need to be tuned in the motor operating range.

The method of Lyons *et al.* (1991) and Lyons and MacMinn (1992) makes use of a set of magnetization curves stored in a multi-dimensional table. Flux-linkage (calculated by integrating the phase voltage) and phase current measurements are made during motor operation in predetermined sensing regions defined over an electrical cycle. This region is where the phase inductance is changing rapidly. For a given flux-linkage and current, the rotor position of the energized phase is estimated from the magnetization curves, Figure 7.1. The method also takes into account mutual coupling effects. Measurements of voltage and current are made simultaneously in each conducting phase, and the rotor position is estimated continuously over the sensing region. The block diagram of the method is depicted in Figure 7.7. This method involves the storage of a significant amount of data and a fast digital signal processor. Jones and Drager (1995) make use of this method for a high-speed starter/generator application, where its performance looks encouraging.

In order to minimize the amount of stored data, Lyons *et al.* (1992) proposed to store only one magnetization curve which represents a reference position θ_{ref} between the unaligned and the aligned positions instead of a multi-dimensional table of a range of positions. The flux-linkage and current of the active phase is measured continuously from the beginning of phase commutation. The current is used as an index to look up the reference flux-linkage in the table. The block diagram of the method is presented in Figure 7.8. Initially the measured flux-linkage ψ_i is smaller than the flux-linkage reference ψ_{ref}, but at some point ψ_i starts to exceed ψ_{ref}. At this time, it is assumed that the rotor position has just passed the reference rotor position θ_{ref}. The comparison between the reference flux-linkage and the measured flux-linkage is done continuously until the reference value is reached. Some extra factors are added in the calculation of rotor position to include mutual coupling coefficients. This method is quite similar

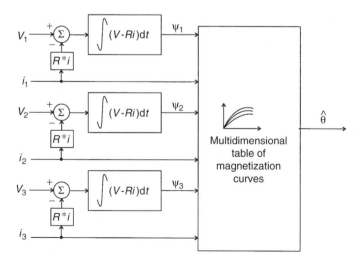

Fig. 7.7 Block diagram of Lyons' method for a three-phase machine.

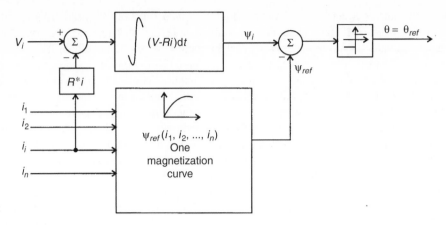

Fig. 7.8 Block diagram of optional Lyons' method.

to Hedlund's approach. The main difference is that Lyons makes use of a vector of a magnetization curve rather than an inductance reference with compensation factors. Lyons and MacMinn (1992b) also proposed to make sure that the rotor position estimator tracks the rotor position by verifying that the phase inductance or flux-linkage falls into a specified tolerance of its theoretical value.

Ray (1995) and Ray and Al-Bahadly (1994), like Lyons, make use of only one magnetization curve which represents a rotor position between the unaligned and the aligned position; however, he uses an extra vector that represents the variation of $\partial\theta/\partial\psi$ as a function of current. The main difference from Lyons' method is that the calculation of flux-linkage and current to estimate the rotor position is done only once at a specific time after the phase was activated. Therefore, the estimated position is not fixed. Referring to Figure 7.9(a), ψ_{ex} is the flux-linkage which would be expected if the rotor position were identical to the reference position θ_r at the sampling time. The error in position is estimated by multiplying the difference $\psi_m - \psi_{ex} = \Delta\psi$ by the value of $\partial\theta/\partial\psi$ at the measured current i_m. This is expressed by,

$$\Delta\theta = \left[\frac{\partial\theta}{\partial\psi}\right]_{i_m} \cdot \Delta\psi \tag{7.6}$$

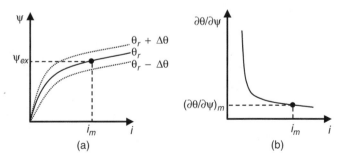

Fig. 7.9 (a) Magnetization curve; (b) $\partial\theta/\partial\psi$ as a function of current.

since the reference position θ_r is known, the true rotor position at the measured instant may be calculated as

$$\theta_m = \theta_r + \Delta\theta. \tag{7.7}$$

Figure 7.9(b) shows the variation of $(\partial\theta/\partial\psi)$ as a function of current for the reference rotor position θ_r. Ray also proposed to estimate the rotor position in the same way at low speed by injecting a test pulse in an unenergized phase.

DiRenzo and Khan (1997) also proposed to use an intermediate magnetization curve for rotor position estimation. The difference is that the intermediate magnetization curve is approximated based on the magnetization curve at the aligned position by

$$\psi(\theta_{ref}, i) = \alpha(i)\psi(\theta_{aligned}, i) \tag{7.8}$$

where $0 \le \alpha(i) \le 1$. The reference position is reached when the estimated flux-linkage is larger than the one calculated by equation (7.8). The dwell angle is fixed and commutation is advanced or retarded by varying α. It is also proposed to obtain the magnetization curve at the aligned position by the drive itself with an initialization routine.

Alternatively, Lyons et al. (1992a) proposed a lumped parameter reluctance network model of the motor, i.e. the model includes a lumped network of stator, rotor and airgap reluctance terms where many are functions of rotor angle θ. Such a model takes into account multi-phase saturation, leakage and mutual coupling effects. Flux-linkage and current for each phase are measured simultaneously. The reluctance terms of the model are determined from these measurements, and are used to estimate rotor position. It is important to note that this method tends to include all the nonlinear effects of the machine which has the effect of requiring a large amount of stored data and computation time.

Lyons and Preston (1996) also suggested using the flux-linkage method to cover all the speed range. The flux-linkage and current of probing pulses are measured in an unenergized phase at low speed, and their values in the energized phase are used at high speed instead.

An example of a fuzzy logic algorithm applied to sensorless control is presented in Cheok and Ertugrul (1996). They claim high performance of the method; however, it seems that it is computationally heavy. A modification of this method to reduce the amount of computation time is done by Xu and Bu (1997). Recently, Mese and Torrey (1997) applied neural networks for rotor position estimation, and high performance of the drive is claimed; however, it needs long off-line time for training the neural network. Both fuzzy logic algorithm and neural network are based on the magnetic characteristics of the machine. Alternatively stated, they are based on the magnetization curves, which are represented by fuzzy logic functions or a neural network respectively.

It should be noted that all these methods need an initial rotor position, therefore an initialization routine is required at standstill.

The main disadvantage of the methods included in this group is the calculation of flux-linkage by integration of the phase voltage. This is because the phase resistance varies strongly with temperature. Its effect is worse at low speeds where integration errors can be large due to long integration periods. Another drawback is the necessity of a significant amount of pre-stored data of the magnetization curves, which increases the amount of operations to be computed. Also, the drive may be limited by the maximum speed of the DSP at high speed. Examples of efforts to reduce computer time in these methods are Hedlund and Lundberg (1991), Lyons *et al.* (1992) and Ray (1995). However, the price to be paid is the reduction of resolution in the estimation of rotor position.

On the other hand, the advantages offered by these methods are: immunity to the effect of saturation, applicable in a wide speed range, the mutual effect can be taken into account, good accuracy and four-quadrant operation of the drive is possible. It seems that these methods are more suitable for medium- and high-speed applications due to the difficulty of calculating flux-linkage at low speed.

It is important to note that the accuracy of the magnetization curves plays a significant role on the accuracy of these methods because they rely on this data. Therefore, special care has to be taken in their measurement.

7.4.4 State observers

This method was proposed by Lumsdaine and Lang (1990). A mathematical model of the complete system, which includes the mechanical load, is run simultaneously with the real system. In such a model, the voltage is considered to be the input and current is considered to be the output. The state variables are flux-linkage, speed and rotor position. The model represented in equations (7.9)–(7.12) assumes constant inertia J and viscous damping B, and an inverse inductance function $H(\theta) = L^{-1}(\theta)$ which is independent of current.[5] The estimated current is compared with the measured current and the error is used as an input to the model to adjust the gains F_ψ (matrix gain), F_ω, and F_θ (vector gain). Figure 7.10 shows a block diagram of the system. It is claimed to have a resolution up to 50 000 pulses per revolution, but based on simulations of a

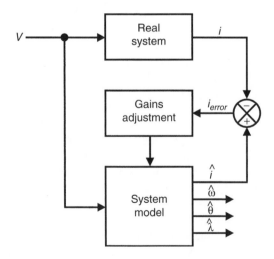

Fig. 7.10 Block diagram of the system based on state observer for rotor position estimation.

---

[5] The use of inductance independent of current implies that saturation is neglected.

linear model of the machine.

$$\frac{d\hat{\psi}}{dt} = -R \cdot H(\hat{\theta}) \cdot \hat{\psi} + v_{ph} + F_\psi(\hat{\psi}, \hat{\theta}) \cdot (\hat{i} - i) \tag{7.9}$$

$$\frac{d\hat{\omega}}{dt} = -\frac{B}{J} \cdot \hat{\omega} - \frac{1}{2 \cdot J} \cdot \hat{\psi}^T \cdot \frac{dH(\hat{\theta})}{d\hat{\theta}} \cdot \hat{\psi} + F_\omega(\hat{\psi}, \hat{\theta}) \cdot (\hat{i} - i) \tag{7.10}$$

$$\frac{d\hat{\theta}}{dt} = \hat{\omega} + F_\theta(\hat{\psi}, \hat{\theta}) \cdot (\hat{i} - i) \tag{7.11}$$

$$\hat{i} = H(\hat{\theta})\hat{\psi}. \tag{7.12}$$

On the other hand, Elmas and la Parra (1993) proposed a reduced order extended Luen-berger type nonlinear observer model. In this case, experimental results are presented, but the motor is operated in its linear region.

Jones and Drager (1998) proposed to use a Kalman filter estimator for absolute rotor position estimation. The method is divided basically into four blocks. The first block, called relative angle estimator patented by Drager *et al.* (1999), estimates the rotor position from each energized phase from the magnetization curves in similar fashion to Lyons. The second block calculates the absolute rotor position based on the position estimation of each energized phase. The third block is the Kalman filter estimator, which estimates rotor position, speed and acceleration based on the posi-tion values given by the second block. The last block, called instantaneous position generation (Jones and Drager, 1999), calculates the instantaneous rotor position based on the position estimation given by the Kalman filter. The last block is necessary because the position estimation obtained by the Kalman filter is not fine enough for phase commutation due to heavy computation. Jones and Drager (1997) also proposed a initialization routine to estimate the initial rotor position. The method consists in injecting probing pulses to all the phases and estimating the initial position value from the magnetization curves.

A sliding mode observer for position estimation has been proposed by Husain *et al.* (1994). This observer is much simpler than Lumsdaine's. The simulation results show good performance of this method. Further study of this method is made by Blaabjerg *et al.* (1996), who implement the method in real time and introduce a saturation function in the sliding observer in order to reduce chattering. McCann and Husain (1997a) present this method but based on flux-linkage observer rather than current observer as in the previous papers. They discuss convergence of the method at the startup and with error in flux-linkage. It is concluded that the sliding mode observer is robust to flux observer errors. McCann (1997) also proposed a hybrid position estimation method using the sliding mode observer and a low resolution position sensor (i.e. one position per energy conversion). The low resolution position sensor is used to correct error in position estimation, particularly at high speed where there is not enough time for the observer to converge. The main advantage of the sliding mode observer is that the complexity of the observer is reduced while its performance is maintained.

Recently, an interesting sensorless method based on current observer has been studied by Acarnley *et al.* (1995) and Ertugrul and Acarnley (1994). Further investiga-tion of this method was carried out by Gallegos-López *et al.* (1999) and Gallegos-López

(1998). The discussion of this method is given in Section 7.7 which includes some simulation and experimental results.

The success of observer methods relies on the accuracy of the mathematical model and the computational power of the hardware. The main disadvantages of these methods are: real time implementation of complex algorithms which require a high speed DSP, a significant amount of stored data and additional extra circuitry which greatly increases the cost, and speed limitations imposed by the DSP.

On the other hand, the advantages are: high resolution in detecting rotor position, high accuracy in estimating rotor position, applicability to the whole speed range, good performance in load torque transients, and applicability at standstill.

7.4.5　Irregularities in stator/rotor poles

This method was proposed by Bartos *et al.* (1993). The method suggests the alteration of the inductance profile by introducing an irregularity such as a notch, in at least one of the stator pole faces and/or at least one of the rotor pole phases. Figure 7.11(a) depicts the irregularities in the stator and rotor poles and their effect in the inductance profile is shown in Figure 7.11(b). If the motor is operated in single pulse or PWM-voltage control mode, the irregularity on the phase inductance shows up in the current waveform and therefore the specific rotor position, where the notch was made, can be detected. One rotor position per energy conversion can be detected.

The main disadvantage is that the inductance profile is affected, and therefore the torque 'profile' produced by the irregular motor is deteriorated and the torque ripple is increased. The extra problem is to make the notches in the stator/rotor poles, which increases the mechanical complexity. Rotor position estimation is not possible at stand-still, and the phase should be energized over the irregularity.

On the other hand, the advantage could be that it is simple to detect one rotor position once the notch has been made. However, this method does not work if the motor is controlled by current regulation.

7.4.6　Current waveform

Sood *et al.* (1995) proposed a method which compensates the commutation angle as a function of the rate of change of the current waveform in an energized phase. If the actual current waveform does not match with the desired current waveform, the

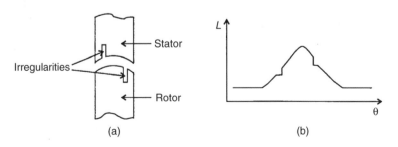

(a)　　　　　　　　　　　　　　　(b)

Fig. 7.11 (a) Stator/rotor irregularities; (b) modified inductance profile.

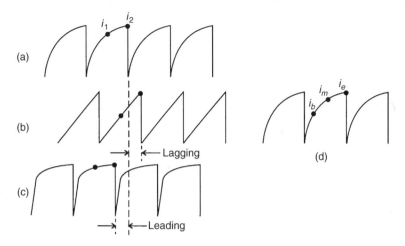

Fig. 7.12 D.c. bus current waveforms: (a) commutation 'in phase' with rotor; (b) commutation lags rotor; (c) commutation leads rotor.

commutation angle is adjusted. The motor is operated in single pulse with a constant dwell angle and no excitation overlap. The feedback signal is the d.c. bus current. Figures 7.12a–c depict a current profile in the d.c. bus current which indicates if the commutation angle is leading or lagging. In (a), (b) and (c) commutation is considered 'in phase', 'lagging' and 'leading' respectively. The current profile is sampled twice, one at the middle (i_1) of the commutation angle and the second just before the phase is turned off (i_2). From this information di/dt is calculated and compared with the desired value. The error indicates if the commutation angle is 'in phase' and it is adjusted accordingly. Marcinkiewicz *et al.* (1995) improved this method introducing the sampling of three current values. The samples are made at $\frac{1}{3}$, $\frac{2}{3}$ of the commutation angle and the last one just before the phase is turned off. This is illustrated in Figure 7.12d. The sampled data is used in equation (7.13) which gives an approximately linear function of commutation angle at constant speed.

$$I_{curve} = 3i_e - 3i_m + i_b. \tag{7.13}$$

The main disadvantages of this method are: the current waveform information of the specific motor is needed for a number of speeds before driving the motor, no specific rotor position is estimated but the motor is stabilized by matching the desired current waveform. The current waveform is speed dependent and therefore a large amount of data may be needed in order to cover the whole motor speed range. The dwell angle must be fixed at 180° electrical. Conversely, the potential advantages are: reasonable stability for speed transients, easy implementation and four-quadrant operation.

A recent method was patented by Lim (1996). Figure 7.13 shows the circuit block diagram for one phase. The phase current waveform is differentiated directly from each phase. The resulting signal di/dt is amplified and compared to a reference value and the output of the comparator is used as a commutation signal (Comm. pulses 1). As an alternative option, the signal di/dt is passed through a low-pass filter in order to introduce a delay. This signal is amplified and compared with a second reference

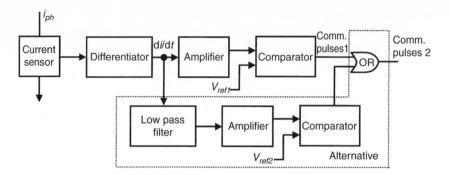

Fig. 7.13 Block diagram of differentiation of current waveform.

value and the output of the second comparator is 'ORred' with the signal of the first comparator to obtain a commutation signal (Comm. pulses 2). The motor may be controlled in voltage control or single pulse, the dwell angle is fixed and it is suggested that the angles can be advanced by adjusting the resistances and capacitances of the low-pass filter.

The main disadvantages of this method are: it is not applicable at standstill or low speed, it is inflexible in advancing the commutation angle and the dwell angle is fixed. It seems that it is applicable to constant speed operation and limited to two-quadrant operation. The advantages of this method are its low cost and simple implementation.

Current gradient and *voltage magnitude sensorless methods* were proposed by Kjaer *et al.* (1994). Both methods detect the specific rotor position where the stator and rotor pole start to overlap. The difference is that the *current gradient* method uses the change in di/dt when the motor is operated with PWM-voltage control or single pulse, and that the *voltage magnitude method* detects the change in the average phase voltage when the motor is operated in constant current regulation. Further study of the *current gradient sensorless method* was done by Gallegos-López *et al.* (1998) and Gallegos-López (1998). The discussion of this method is given in Section 7.6 which includes some experimental results.

Laurent *et al.* (1995) proposed the use of eddy current variation to estimate rotor position when the motor is operated in PWM-voltage control. It is claimed that there is a discontinuity in the current waveform at the switching instant depending on the iron losses. The rotor iron losses are dependent on rotor position, being minimum at the aligned position and maximum at the unaligned position. Therefore, the rotor position can be estimated by an inverse function of the iron losses. The iron losses are extracted from the fundamental of the PWM-voltage and current in the phase. It is also shown that the iron losses depend on the magnetic saturation, therefore only an intermediate position is estimated where magnetic saturation has low effect. The main disadvantage of the method is that the motor cannot be operated in single pulse mode. This means that it is only suitable for low and medium speed.

Alternatively, Watkins (1998) proposed in a recent patent to use the rate of change of the phase current at particular rotor position where the phase current is forced to freewheel in a zero voltage loop (i.e. only one switch is on). He suggested to use the aligned position as a reference, where the rate of change of the phase current must be zero. The phase is turned off close to the aligned position, at this time the current freewheels through both diodes, and after a period of time, one switch is closed to force a zero voltage loop for a short period of time. At this time di/dt is monitored. The magnitude and polarity of di/dt represents a rotor position shift from the aligned position and whether it is advanced or delayed. The aligned position is chosen as a reference because di/dt is always zero regardless of the flux that is in the machine. Another position could be the unaligned position, where di/dt is also zero. This method estimates one position per energy conversion. The advantage is that it is easy to implement and it works for the whole speed range; however, it is important to note that the de-energizing period of the phase has to be extended, which may result in a decrease of motor efficiency. The disadvantages of this method are that stored data may be required and it is not applicable at standstill.

7.5   Unenergized phase methods

Most of the methods of this group measure the phase inductance $L(\theta)$ in an unenergized phase, which is used to estimate the rotor position. In other words, position estimation is made when the phase is not needed for torque production. Generally, these methods use the injection of a low level chopping current waveform in unexcited phases. The advantages of these methods are that they minimize the effect of the back-EMF and eliminate the effect of magnetic saturation (problems found in many sensorless methods that use the active phase), because the position is estimated in the linear region of the machine. The current rise or fall time of probing pulses may be measured in an unexcited phase for rotor position estimation, in a similar manner to the chopping waveform for an active phase. The method of injecting probing pulses has been investigated and improved by many researchers. Other sensorless methods, included in this group, use modulation techniques similar to those used in communication systems, or a modulated resonant frequency to calculate the phase inductance and hence the rotor position. The use of low level regenerative current in an unexcited phase, and the use of mutual induced voltage in an unexcited phase for indication of rotor position are methods that also belong to this group.

7.5.1   Active probing

Acarnley *et al.* (1985) and Hill and Acarnley (1985) suggested, in a third method, to use a low chopping current waveform in a non-torque productive phase. The probing pulses are injected at high frequency from the main power circuit. The current peak of the probing pulses is relatively small (usually 10% of the rated current) so that any negative produced torque can be neglected. The pulses are normally of fixed duration at frequencies in the range of 4 to 20 kHz, depending on the phase inductance.

Assuming that the pulse period is short enough, the peak current is small, and the phase voltage[6] can be approximated as:

$$v \approx L(\theta)\frac{i_{peak}}{\Delta T}. \tag{7.14}$$

---

[6] The voltage drop ($R \cdot i$) and the back-EMF (e) are neglected. However, equation (7.14) is still valid because $R \cdot i + e \ll L(t)(i/\Delta t)$.

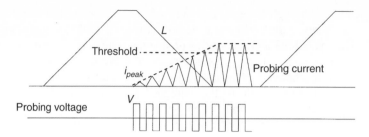

Fig. 7.14 Probing pulses in an unexcited phase.

Rearranging equation (7.14), it is found that the peak current is inversely proportional to the instantaneous value of the phase inductance

$$i_{peak} \approx \frac{v \cdot \Delta T}{L(\theta)}. \tag{7.15}$$

Making use of equation (7.15), the phase commutation is found simply by comparing i_{peak} with a threshold value. The commutation angle can be advanced or retarded by reducing or increasing such a threshold value respectively. An alternative method is to measure the inductance profile continuously for use as an index in a look-up table. Another approach is to use an inverse function (7.16) for calculation of rotor position. Figure 7.14 depicts typical probing pulses in an unexcited phase. The probing pulses normally are injected in the phase next to be energized.

$$\hat{\theta} = G^{-1}(\hat{L}). \tag{7.16}$$

Harris and Lang (1990) studied the effect of eddy currents, inverter switching noise, magnetic coupling and digital quantization for this method. It is concluded that these effects have a big impact on the performance of this method, and some suggestions are given to minimize them.

Alternatively, Mvungi *et al.* (1990) and Mvungi and Stephenson (1991) claim that the flux-linkage at constant current has better sensitivity than the current at constant flux-linkage. Therefore, he proposed to use probing pulses of constant peak current (variable frequency) in an unenergized phase. In this case, the rotor position is estimated by observing the magnitude of the flux-linkage of the probing pulses.

MacMinn *et al.* (1988), MacMinn and Roemer (1988) and MacMinn *et al.* (1990) suggested estimating the rotor position in two unenergized phases simultaneously to avoid corruption in the estimated position due to switching noise or mutual coupling. If two phases are not unenergized during the sampling period, it is suggested that the extrapolated rotor position should be used instead of the instantaneous estimated value. They also suggested monitoring the variation of the supply voltage for more accurate rotor position estimation.

A similar method was proposed by Egan *et al.* (1991), but they proposed to inject the probing pulses at different frequency than the switching frequency of the active phase in order to avoid the effect of mutual coupling.

Van Sistine (1996b) suggested comparing the peak current of the probing pulses in two unenergized phases. When both peak currents are equal the next phase in the sequence is energized. Hedlund and Lundberg (1989) proposed improvements in inductance measurement by introducing a compensation factor which takes into account the mutual coupling effect. They measure $L_a(\theta, i_b)$ of phase 'a' in the presence of current i_b in the active phase and adjust this according to the equation:

$$L_a(\theta, 0) = (1 + c \cdot i_b)L_a(\theta, i_b). \tag{7.17}$$

McCann (1997b) and McCann (1999) proposed to use probing pulses in an unenergized coil phase to estimate rotor position similar to Acarnley's approach. However, McCann suggests having two set of coils per pole (four per phase) and therefore two asymmetric bridges per phase. One set of coils is energized to drive the motor while probing pulses are injected in the other set of coils to estimate rotor position. The main advantage of this method is that the time for probing pulses is not limited at high speed because the set of coils is independent. On the other hand, the disadvantage is that two asymmetric bridges per phase are required.

Recently, Blackburn (1998) patented a method based also on probing pulses in an unenergized phase. The improvement that he suggests is to compensate for variations in the phase inductance. The compensation consists in holding the magnitude of the first probing pulse (minimum peak) after the phase is unenergized and the last probing pulse before the phase is energized (maximum peak). The threshold is not a fixed reference value but it is obtained as a function of the minimum and maximum peak, and the position is estimated when the peak of the probing pulses exceeds the threshold.

Vitunic (1999) proposed injecting pulses in an unenergized phase close to the unaligned position. The peak current of the first probing pulse is stored, then subsequent peak pulses are compared to a predetermined threshold which is used to determine if the rotor is stationary. Once the peak current of the probing pulses exceeds the threshold, the subsequent probing pulses are compared with the initial stored peak current. When the peak current does not exceed the stored value, commutation to the next phase is carried out. The main difference from previous methods is that the threshold is not used as a reference to commutate the next phase but to make sure that the rotor is moving. The main advantage of this method is that the effect of variations in the phase inductance from phase to phase and voltage supply is avoided.

These methods possess the following disadvantages. It is usually difficult to implement them at high speed because current flows in a phase for almost the whole electrical cycle, hence there is little time for probing pulses. For this reason, the speed range is limited up to medium speed. The current in the active phase induces voltage in the unenergized phases which strongly distorts the probing pulses, therefore these methods are very sensitive to mutual coupling. There is the necessity of knowing the specific phase inductance as a function of rotor position. It is difficult to estimate the rotor position close to the unaligned position due to the flatness of the phase inductance in this region. The injection of probing pulses deteriorates the overall drive efficiency.

The main advantages are: four-quadrant operation is possible, the probing pulses are injected by the main converter itself and no extra circuit is required, it is applicable at standstill, unique rotor position information can be obtained if two phases are probed, the effect of the back-EMF is minimized, and the saturation effect is avoided due to the low current level of the probing pulses.

It is important to note that the initial rotor position at standstill can be estimated by injecting probing pulses in all phases. The position can be estimated in different ways from the current and/or the voltage of the probing pulses. An example of using the rate of change of the current of the probing pulses is presented by Van Sistine (1996a).

7.5.2 Modulated signal injection

Ehsani *et al.* (1992), Ehsani *et al.* (1994) and Ehsani (1991, 1994, 1995) proposed to measure the phase inductance in an unenergized phase by modulation techniques like those used in communication systems (frequency, amplitude and phase modulation). In the frequency injection method (Ehsani *et al.*, 1992; Ehsani, 1991), the phase winding of an unenergized phase is connected to an oscillator shown in Figure 7.15. The oscillator is designed such that the frequency f is inversely proportional to the phase inductance, equation (7.18).

$$L(\theta) = \frac{1}{k \cdot f_{mod}}. \tag{7.18}$$

The probing phase should be connected to the external frequency modulator/demodulator circuit as shown in Figure 7.16. The method requires a multiplexer for connection and disconnection of the modulator from the inverter. In Ehsani *et al.* (1994) and Ehsani (1994, 1995), an alternative is suggested for the frequency injection method. This time, a low level sinusoidal voltage with fixed frequency and amplitude is applied to an unexcited phase from an external oscillator via a resistance. The phase displacement, equation (7.19), and the amplitude, equation (7.21), between voltage and current vary as a function of the phase inductance. Therefore the inductance can be measured by detecting the change in either phase displacement, equation (7.20), or amplitude, equation (7.22). Ehsani claims that phase modulation and amplitude modulation are better for position estimation at low inductance and high inductance respectively. Therefore, a combination of both is suggested in order to obtain better

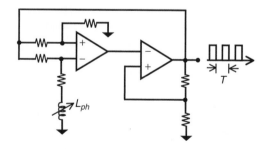

Fig. 7.15 Oscillator.

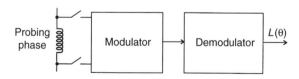

Fig. 7.16 Block diagram of the modulated signal technique.

rotor position estimation and robustness of the method over the whole electrical cycle.

$$\alpha = \tan^{-1} \frac{\omega \cdot L(\theta)}{R} \qquad (7.19)$$

$$L(\theta) = \frac{R \cdot \tan \alpha}{\omega} \qquad (7.20)$$

$$I = \frac{v \cdot \sin(\omega \cdot t - \alpha)}{\sqrt{R^2 + (\omega \cdot L(\theta))^2}} \qquad (7.21)$$

$$L(\theta) = \frac{1}{\omega} \sqrt{\frac{v^2}{I^2} - R^2}. \qquad (7.22)$$

The main disadvantages of the modulation techniques are: a multiplexer is required to connect the external modulator/demodulator to the probing phase, the sensing circuit needs to be isolated from the power converter, the test signal is susceptible to corruption due to mutual coupling effects, the specific phase inductance should be known, an external modulator/demodulator circuit is required, hence increasing the analogue circuitry, and the speed range is limited up to medium speed where there is sufficient zero current interval. On the other hand, these methods may offer four-quadrant operation, the effect of the back-EMF is minimized and the saturation effect is avoided, the inductance is measured rather than the impedance, and the position may be estimated with reasonably good accuracy. It is applicable at standstill.

The method proposed by Harris *et al.* (1993) includes a resonant tank (*RLC*) connected to an unenergized phase, where *L* is the phase inductance. A low power signal is injected into the tank of an unenergized phase similar to Ehsani's method. The resonant frequency characteristic of the tank varies between maximum and minimum values because of variation of the phase inductance. Therefore, it is possible to use the variation of the resonant frequency to estimate the rotor position. The low power resonant circuit is connected to the machine by coupling capacitors and therefore no multiplexer is required. Laurent *et al.* (1993) show that the use of resonant tank (*RLC*) increases the accuracy of inductance measurement compared with an *RL* circuit.

7.5.3 Regenerative current

A recent method proposed by Van Sistine *et al.* (1996) uses the regenerative current in an unenergized phase. The phase is turned off before the aligned position, but approximately 15° (mech.) after the aligned position, the upper and the lower switches of the power converter are turned on again. The lower switch is kept on while the upper switch is chopped at fixed frequency in order to maintain the current about 4% of the rated current, hence no significant negative torque is generated. The peak current is monitored and when it exceeds a reference value the next phase in the sequence is energized. Figure 7.17 illustrates the principle of the method.

The disadvantages of this method are: it may not be applicable to high speed, it may generate considerable negative torque and therefore the overall efficiency of the drive is deteriorated, it is not applicable at standstill, and it may limit the possible commutation angles. The advantages could be its easy implementation (it seems to be applicable to constant speed operation) and that precise inductance data is not required.

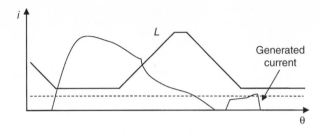

Fig. 7.17 Regenerative current in an unenergized phase.

7.5.4 Mutually induced systems

Austermann (1993) proposed to use the mutually induced voltage in an unenergized phase produced by the current in the conducting phases. The induced voltage is expressed as a function of the mutual flux-linkage ψ_{ml} by equation (7.23) which varies significantly with rotor position.

$$v_{ind} = \frac{d\psi_{ml}(i_{active},\theta)}{dt} = \left[\frac{\partial\psi_{ml}}{\partial i_{active}}\right]_{\theta=cst} \cdot \frac{di_{active}}{dt} + \left[\frac{\partial\psi_{ml}}{\partial\theta}\right]_{i=cst} \cdot \frac{d\theta}{dt}. \tag{7.23}$$

He claims that the induced voltage passes through zero at a known position determined by the motor geometry, and therefore a position estimation is obtained. Husain and Ehsani (1994b) used practically the same method. They suggest splitting the induced voltage in two values, one for the on time and the other for the off time of the PWM of the active phase. This means that sampling and hold of the induced voltage has to be synchronized with the PWM switching frequency. The commutation instants are obtained by comparing the induced voltage with a threshold. Bin-Yen *et al.* (1997) simplify this method by proposing to rectify the induced voltage. In this case, there is no need of sampling and hold and synchronization with the PWM switching frequency.

Horst (1997) proposed to observe the induced current in an unenergized phase when the coils are in parallel or series–parallel in the case of a 12/8 three-phase machine. He claims that the current flowing in the closed unenergized phase path has a pronounced indentation, which is representative of rotor position, due to parameter variation in the machine from phase to phase such as airgap, phase inductance and phase resistance. The main disadvantage of this method is that at least an extra pair of leads and one extra current sensor per phase are necessary.

The method seems to be limited to systems using constant current regulation because the mutual voltage induced depends on the level of the current in the excited phase and therefore the current should be constant over the conduction period of the active phase (this means that current profiling is not allowed). It is important to note that the method may be corrupted by noise in the system, because the ratio between induced voltage and system noise is small. This is the main disadvantage of this method. Furthermore the speed range is limited up to base speed, where there is enough zero current period to observe the induced voltage. The possible advantage is that the method estimates the rotor position by the direct measurement of an internal signal, which is available without the injection of any diagnostic pulses.

7.6   Current gradient sensorless method

The *current gradient sensorless method* (CGSM) makes use of the change of the slope of the phase current waveform (di/dt) to estimate one rotor position per energy conversion. The position that is estimated corresponds to the beginning of overlap between the rotor and the stator poles. The motor must be controlled in PWM-voltage control below its based speed and single pulse above its base speed.[7] This method was originally proposed by Kjær *et al.* (1994b) and further development was made by Gallegos-López *et al.* (1998) and Gallegos-López (1998). It belongs to the subgroup *current waveform* in the group *energized phase* of sensorless control methods discussed in Section 7.4.6. The main advantage of this method is that no previous knowledge of the motor parameters is required, except the pole geometry.

The principle of this method is very simple. For motoring, it defines that *the slope of the phase current is always larger for $\theta < \theta_o$ than for $\theta > \theta_o$* (see Figure 7.18, top). For generating, it says that *the slope of the generated phase current is always larger for $\theta < \theta_b$ than for $\theta > \theta_b$* (see Figure 7.18, bottom). Note that the change in di/dt is caused by the change in the phase inductance. Therefore, the specific positions θ_o for motoring and θ_b for generating can be detected by simply observing the change of di/dt in the phase current waveform.

The fully analogue detection stage is shown in Figure 7.19. It consists of a current sensor, two low-pass filters, a differentiator, and a zero-crossing detector. The low-pass filters are used to eliminate the PWM switching frequency and possible noise.

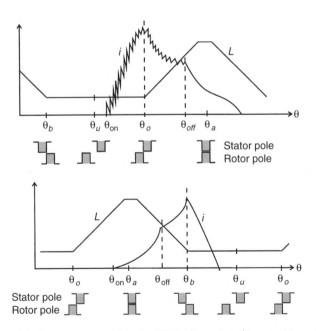

Fig. 7.18 Upper: typical phase current waveform for PWM voltage control in motoring mode. Lower: typical phase current waveform in generating mode. (In Gallegos-López *et al.* 1998. © IEEE 1998).

---

[7] The switched reluctance machine is usually operated in single pulse above its base speed.

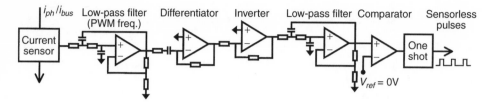

Fig. 7.19 Detection stage for di/dt = 0. (In Gallegos-López *et al.* 1998. © IEEE 1998).

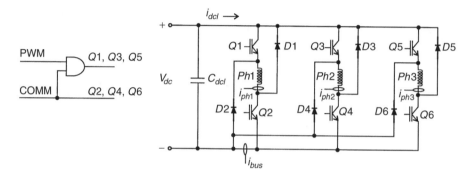

Fig. 7.20 SRM inverter with split lower rail. (In Gallegos-López *et al.* 1998. © IEEE 1998).

The differentiator is used to obtain the derivative of the current waveform di/dt, and the zero-crossing detector gives a pulse when di/dt is zero. Note that if the drive is going to be operated in single pulse mode (i.e. no PWM) the low-pass filters are not necessary and the number of components in the detection stage is reduced significantly.

The feedback signal to detect the rotor position can be either the current waveform of each phase or the current waveform of the lower controlled switch bus (i_{bus}), which contains the same information as the phase currents (see Figure 7.20). The former requires one current sensor and detection stage per phase. In contrast, the latter requires only one current sensor and detection stage which significantly reduces the circuitry.

Some experimental results are discussed in the following paragraphs for sensorless operation of a low-power motor. Figure 7.21 shows one phase current (i_{ph1}), phase lower controlled switch bus current (i_{bus}), the *current gradient position estimation* (CGPE) pulses obtained from i_{bus}, and the decoded pulses (DP) for phase one measured at 1763 rpm in single pulse with $\theta_{on} = 50°$, $\theta_{off} = 80°$ (no excitation overlap). Clearly, the correct rotor position θ_o is detected.

Figure 7.22 illustrates the case of excitation overlap showing i_{ph1}, i_{bus}, CGPE and DP for phase one measured at 1820 rpm, with $\theta_{on} = 50°$, $\theta_{off} = 84°$. Note that this time, two sensorless pulses appear per stroke, the first one is erroneous (when the previous phase is turned off) and the second one gives θ_o. A simple digital logic circuit is implemented to neglect the first pulse when there is excitation overlap, so DP becomes the decoded signal for phase one.

Figure 7.23 depicts the estimated position $\hat{\theta}$ and the position given by a 1024-line encoder θ in steady-state with no load measured at 2304 rpm, with $\theta_{on} = 45°$, $\theta_{off} = 80°$. Clearly the position signals show good agreement. It can be observed that

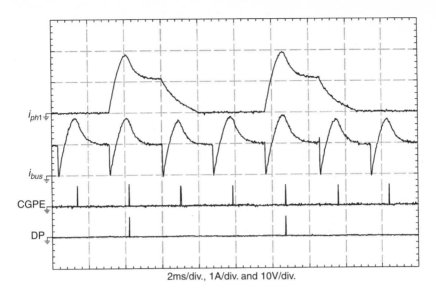

2ms/div., 1A/div. and 10V/div.

Fig. 7.21 Single-pulse current waveforms, $\theta_{on} = 50°$, $\theta_{off} = 80°$ (no excitation overlap). (In Gallegos-López *et al.* 1998. © IEEE 1998).

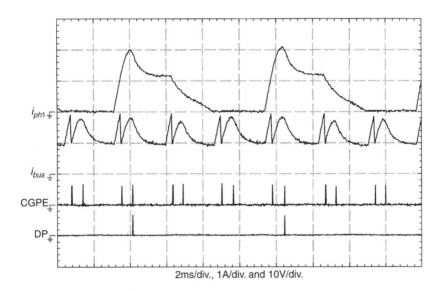

2ms/div., 1A/div. and 10V/div.

Fig. 7.22 Single-pulse current waveforms, $\theta_{on} = 50°$, $\theta_{off} = 84°$ (excitation overlap). (In Gallegos-López *et al.* 1998. © IEEE 1998).

the estimated position is leading by approximately 2.4° (mech.). However, it should be noted that at higher speed the estimated position may lag because the low-pass filters in the detection stage may impose a delay. It is worth noting that $\hat{\theta}$ is steady, clean and it does not present any oscillations.

CGSM cannot be applied at standstill, thus it needs a startup routine. Feedforward may be used, which consists of applying a ramp of frequency to the motor in open-loop

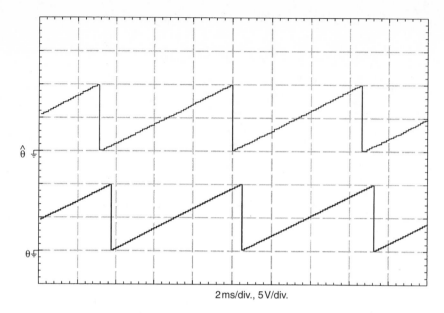

2 ms/div., 5 V/div.

Fig. 7.23 CGSM position estimation at 2304 rpm, single pulse, $\theta_{on} = 45°$, $\theta_{off} = 80°$. (In Gallegos-López *et al.* 1998. © IEEE 1998).

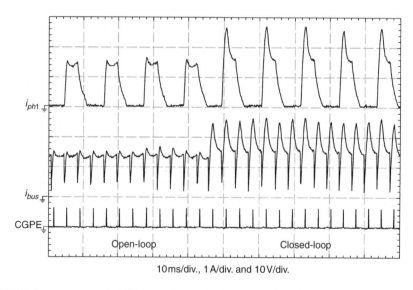

10 ms/div., 1 A/div. and 10 V/div.

Fig. 7.24 Takeover at a speed of 1339 rpm, $\theta_{on} = 50°$, $\theta_{off} = 80°$. (In Gallegos-López *et al.* 1998. © IEEE 1998).

similar to a stepper motor. In this way, the motor is accelerated up to the speed where sensorless pulses can be detected (called takeover speed). Once the takeover speed has been reached, the CGPE pulses can be used to commutate the phases in closed-loop sensorless mode. Figure 7.24 shows i_{ph1}, i_{bus} and CGPE measured during the transition from open-loop to closed-loop true sensorless at a takeover speed of 1339 rpm with,

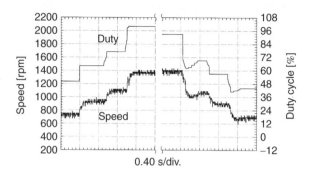

Fig. 7.25 Measured sensorless speed transients. (In Gallegos-López *et al*. 1998. © IEEE 1998).

$\theta_{on} = 50°$, $\theta_{off} = 80°$. The difference in the current waveform is due to difference in open-loop and closed-loop commutation angles.

Figure 7.25 depicts a series of speed transients, which demonstrates that closed-loop CGSM could be acceptable in many low-cost variable-speed applications.

The advantages of this method are: no *a priori* knowledge of inductance profile or magnetization curves are required, it identifies one specific position per energy conversion, where the stator and rotor poles start to overlap, and the commutation angles can be set freely with the condition of $\theta_{on} < \theta_o$, its implementation is simple with a minimum number of extra components, the position can be estimated in a multiphase machine from only one current sensor i_{bus} even with excitation overlap, and finally no extra computation, control requirements or compensation factors are needed.

On the other hand, it suffers from the following disadvantages. It is not applicable at standstill, thus it needs a startup procedure, it is not suitable at very low speed, and it does not permit current regulation for torque smoothing or reduction of acoustic noise.

CGSM is suitable for medium and high speed, given that the peak in the current waveform becomes more prominent with increased speed. Furthermore, the low resolution is adequate at high speed because the speed is approximately constant. Possible applications for this method could be fans, pumps and even domestic appliances. It is also important to note that CGSM is comparable to the back-EMF position estimation method for a brushless d.c. motor in performance and cost.

7.7 Sensorless method based on flux-linkage and current observers

This section describes the principle of a high resolution sensorless method for a switched reluctance motor drive, using either flux-linkage or current to correct for errors in rotor position estimation. The algorithm is capable of estimating the rotor position accurately with fine resolution. This sensorless method makes full use of the nonlinear magnetic characteristics of the switched reluctance machine through correlation of flux-linkage (ψ), current (i) and rotor position (θ), which is shown in Figure 7.26.[8] Therefore, all the nonlinearities of $\theta(i, \psi)$ are taken into account.

---

[8] Note that this data is the same as the one presented in Figure 7.1, but plotted in a different way.

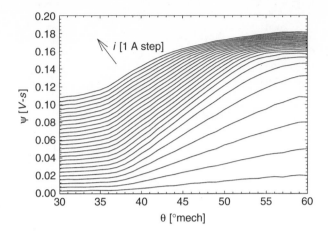

Fig. 7.26 Measured magnetization curves. (In Gallegos-López *et al.* 1999. © IEEE 1999).

The rotor position estimation method using a current correction model (current observer) was initially proposed by Acarnley *et al.* (1995) and Ertugrul and Acarnley (1994), who demonstrated that the algorithm works for the whole speed range and conduction angles.

Further development of the algorithm was made by Gallegos-López *et al.* (1999) and Gallegos-López (1998), who proposed a simpler variation of the algorithm, but with no loss in accuracy, leading to a reduction in real-time computations. Furthermore, a criterion is proposed to choose the most suitable phase for position estimation from all conducting phases.

The algorithm is discussed as follows with the help of its block diagram shown in Figure 7.27. The block 'Position Prediction' extrapolates the next position θ_p from previous and actual position. The block 'Choose Best Phase' identifies the most suitable phase N_b for position estimation from all conducting phases. The 'Integrator block' predicts the flux-linkage ψ_p from the integration of the phase voltage. The block called *Stage I* obtains $\Delta\theta$ from the best phase. The block 'Position Correction' corrects the initial predicted position θ_p in order to obtain the final estimated position θ_e. The block called *Stage II* estimates the flux-linkage ψ_c for each phase for the next step in the integrator block.

For *Stage I*, it is proposed to use either equation (7.24) (*flux-linkage observer*) or equation (7.25) (*current observer*) to calculate the error in position $\Delta\theta$, and equation (7.26) for *Stage II*. Furthermore, it was demonstrated that both flux-linkage observer and current observer are equally sensitive to errors in the predicted flux-linkage for position estimation. Therefore, there is no real advantage of implementing either one.

$$\Delta\theta = \left[\frac{\partial\theta}{\partial\psi} \right]_{i=cst} \cdot \Delta\psi. \tag{7.24}$$

$$\Delta\theta = \left[\frac{\partial\theta}{\partial i} \right]_{\psi=cst} \cdot \Delta i. \tag{7.25}$$

$$\psi_e = \psi_p. \tag{7.26}$$

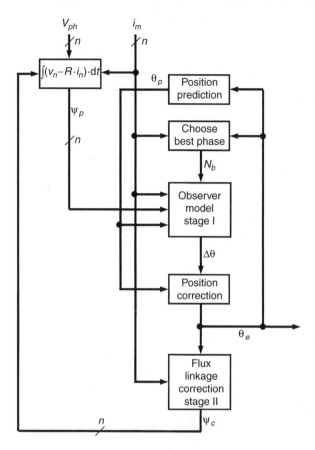

Fig. 7.27 General flow diagram of the proposed position estimation algorithm. v_{ph} = phase voltage, i_m = measured current, θ_p = predicted position, N_b = the best phase for position estimation, $\Delta\theta$ = position error, θ_e = estimated position, ψ_p = predicted flux-linkage, ψ_c = corrected flux-linkage for the next integration step. (In Gallegos-López *et al.* 1999. © IEEE 1999).

After obtaining the error in position $\Delta\theta_n$ for each phase in *Stage I*, the initial predicted position θ_p can be corrected by using one of the following options:

1. The errors in position for each phase ($\Delta\theta_n$) are averaged in order to obtain a single error position value ($\Delta\theta$); it gives,

$$\Delta\theta = \frac{\left[\Delta\theta_1 + \Delta\theta_2 + \Delta\theta_3 + \cdots + \Delta\theta_n\right]}{n}. \tag{7.27}$$

2. The errors in position for each phase ($\Delta\theta_n$) are weighted to obtain $\Delta\theta$; it results in,

$$\Delta\theta = \left[k_1 \cdot \Delta\theta_1 + k_2 \cdot \Delta\theta_2 + k_3 \cdot \Delta\theta_3 + \cdots + k_n \cdot \Delta\theta_n\right]. \tag{7.28}$$

3. $\Delta\theta$ is equal to $\Delta\theta_b$, i.e. the position is corrected from the phase which can give more precise position information.

$$\Delta\theta = \Delta\theta_b, \tag{7.29}$$

where n = number of conducting phases, k = weighting factor and subscript b = *the best phase*. It is important to note that option 1 and 3 are special cases of option 2.

It is known that the position estimation is less accurate ($\Delta\theta_n$ is bigger) in the region close to the unaligned (θ_u) and the aligned (θ_a) position due to the small changes in $[\partial\psi/\partial\theta]_{i=cst}$ (see Figure 7.26) (Lyons *et al.*, 1991; Lyons and MacMinn, 1992b).

The first two options combine the error of all phases that have current into a single position correction factor. Therefore, it is clear that equation (7.27) is more likely to obtain a wrong $\Delta\theta$ when one phase, among all, is close to either θ_u or θ_a. In contrast, equation (7.28) will obtain a more precise $\Delta\theta$ when one of the phases is close to θ_u or θ_a.

The third option estimates the position from *the best phase* among all conducting phases using equation (7.29). Therefore it avoids errors around θ_u and θ_a. A criterion for option 3 is proposed in Gallegos-López *et al.* (1999) and Gallegos-López (1998). For the specific model chosen equation (7.24), it is necessary to minimize the error $\Delta\theta$, hence $[\partial\theta/\partial\psi]_{i=cst}$ should be minimum, or its inverse $[\partial\psi/\partial\theta]_{i=cst}$ maximum.

Figure 7.28 depicts the normalized version of $[\partial\psi/\partial\theta]_{i=cst}$, which shows that the maximum resolution (minimum error $\Delta\theta$) is obtained around 40° and 18 A for the specific 8/6 four-phase machine used. The minimum resolution is around θ_u and θ_a. Close to θ_a the resolution decreases as the current increases. This information indicates which is *the best phase* to give effective position error. Also, it could be used to represent the k_n factors for option 2. Option 3 has the advantage of leading to the most precise position estimation with minimum calculation when there is current flowing in more than one phase.

The results of Figure 7.29 were obtained by simulation using equation (7.24) with $\theta_{on} = 30°$, $\theta_{off} = 55°$, at 716 rpm and 8 A. (a) and (b) depict the error in estimated position for option 1 and option 3 respectively. From the results, it can be observed that the error with option 1 is around 1°, −2° while with option 3 it is around ±0.5°. It is clear that option 3 leads to a smaller position error than option 1.

Figure 7.30(a) depicts the normalized version of $[\partial\psi/\partial\theta]_{i=cst}$ for phase 1 (solid line) and phase 4 (dashed line), note that their intersection indicates that the position

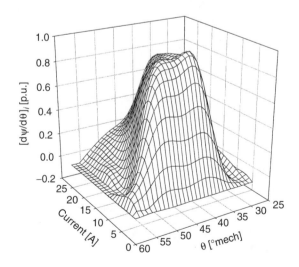

Fig. 7.28 $\partial\psi/\partial\theta\,|_{i=const}$ normalized. $\theta_u = 30°$, $\theta_a = 60°$. (In Gallegos-López *et al.* 1999. © IEEE 1999).

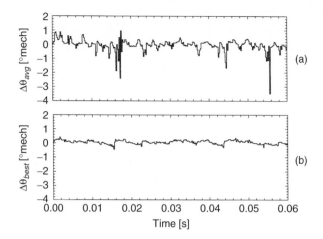

Fig. 7.29 Comparison between error using the average and best phase method: (a) $\Delta\theta_{avg}$; (b) $\Delta\theta_{best}$. Note: $\Delta\theta$ is the difference between the real and estimated positions. (In Gallegos-López *et al.* 1999. © IEEE 1999).

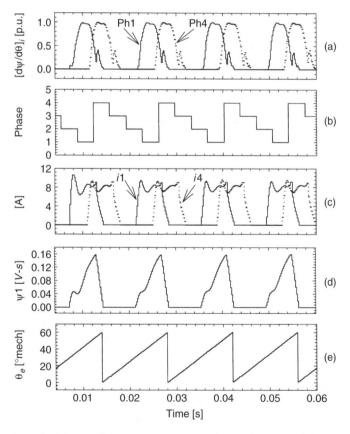

Fig. 7.30 Flux observer model using the best phase for position estimation: (a) $\partial\psi/\partial\theta\,|_{i=const}$ for phases 1 and 4; (b) phase from which the position is estimated; (c) current in phase 1 (solid line) and 4 (dashed line); (d) flux-linkage for phase 1; (e) estimated position. (In Gallegos-López *et al.* 1999. © IEEE 1999).

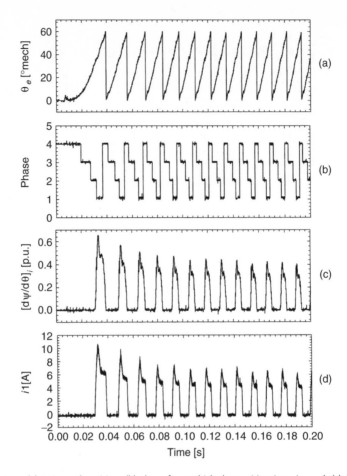

Fig. 7.31 Startup: (a) estimated position; (b) phase from which the position is estimated; (c) $\partial\psi/\partial\theta\,|_{i=const}$ for phase 1; (d) current in phase 1. (In Gallegos-López *et al.* 1999. © IEEE 1999).

estimation should be swapped from phase 1 to phase 4, which results in graph (b) showing *the best phase* (N_b) for position estimation. (c) depicts the current for phase 1 and phase 4, which have an excitation overlap of $10°$, (d) depicts the flux-linkage for phase 1 and (e) shows the estimated position.

Some experimental results are presented in Figures 7.31 and 7.32. Figure 7.31 illustrates the startup sequence with a duty cycle of 60%, the estimated position is depicted in (a), a small ripple is observed due to the swapping of position estimation from phase to phase. (b) shows the phase from which the position is estimated, (c) and (d) depict $[\partial\psi/\partial\theta]_{i=cst}$ and current for phase 1 respectively. Note that at higher current the resolution of position estimation is higher.

An example under PI current regulation is depicted in Figure 7.32 when a step in current reference is applied. (a) shows the estimated position and it is observed when the motor is accelerated. (b) depicts $[\partial\psi/\partial\theta]_{i=cst}$ for phase 1 (solid line) and 4 (dashed line), note that its value is increased proportional to the phase current which is depicted in (c) for phase 1.

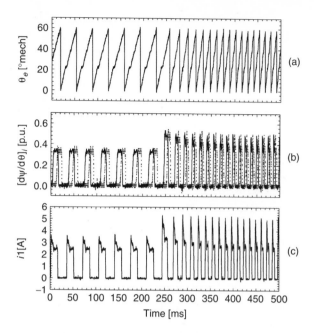

Fig. 7.32 Step in current, $\theta_{on} = 30°$, $\theta_{off} = 55°$: (a) estimated position; (b) $\partial\psi/\partial\theta \mid_{i=const}$ for phases 1 and 4; (c) current in phase 1. (In Gallegos-López *et al.* 1999. © IEEE 1999).

The main advantage of this algorithm is that instantaneous rotor position can be estimated with minimum real-time computations. This is possible because the algorithm makes use of a single phase (*the best phase*) for position estimation based on a simple model of the machine. The advantages are: high resolution of position estimation, the algorithm leads to minimum mathematical operation without decreasing its accuracy, it identifies the most accurate phase for position estimation from all conducting phases, it is applicable either using current and voltage from energized phase or unenergized phase,[9] it is suitable for wide speed range,[10] it is applicable at standstill with an initialization routine, it is applicable to four-quadrant dynamic operation of the drive and it permits load torque transients, it does not compromise the performance of the motor and it allows any motor control.[11]

On the other hand, the main disadvantage of the algorithm is the prediction of flux-linkage which is calculated by the integration of the phase voltage. Other drawbacks are: the algorithm is limited at high speed by the speed of the DSP to cope with the mathematical operations, pre-stored accurate data of magnetization curves is needed.

It is important to note two further points: (1) the algorithm is sensitive to any error in the magnetization curves; (2) the algorithm is not capable of correcting errors in the magnetization curves.[12] The electronics needed may be a significant part of the total cost of the drive, and it suffers from the prediction of flux-linkage by the integration of

---

[9] This is possible by injecting high-frequency probing pulses in unenergized phases.

[10] High speed is limited by the speed of the DSP.

[11] It means that current control, PWM-voltage control, soft and hard chopping, single pulse operation and current profiling are possible, and the commutation angles can be set freely.

[12] In other words, any error in the magnetization curves will cause an error in the position estimation.

phase voltage. The main reasons are: (1) the phase resistance changes in proportion to the motor temperature; (2) uncertainties in the calculation of v_{ph} due to voltage drop in leads from the power converter to the motor and power switches; (3) long integration time; (4) error in current measurements due to noise in the system.[13]

This method offers estimation of instantaneous rotor position with good accuracy. Therefore, it has potential for servo application. The accuracy may be enhanced if the current and voltage of probing pulses are used at low speed and the phase current and phase voltage at high speed.

7.8 Summary

This chapter has reviewed the existing methods for rotor position estimation for switched reluctance motors. It is worthy to note that while most of the methods have been discussed for the case of motoring, they may also be applicable to the case of generating.

Switched reluctance motors possess a unique relationship between flux-linkage, phase current and rotor position. This makes it possible to estimate the rotor position indirectly. Hence, the fundamental principle of most of the sensorless methods is based on the detection of the phase inductance or flux-linkage. In other words, the sensorless methods make use of the phase inductance variation in one way or another.

Until now, there is no method suitable for a wide speed range over which the torque capability of the motor is not limited, i.e. a method that replaces fully the mechanical sensor attached to the motor shaft.

On the other hand, it is evident that extra circuitry and computing time are required to implement any sensorless scheme. The methods that estimate instantaneous rotor position are likely to need a significant amount of stored data, computing time and extra circuitry. In contrast, the methods that estimate few rotor positions per revolution or compensate the commutation somehow are relatively simple to implement and they do not need much computing time or extra circuitry. There are several schemes that do not even need any stored data.

The selection of the sensorless method may differ according to the application. For instance, methods which use a probing signal in an unenergized phase are more suitable for low speed. These methods are not applicable at high speed. At high speed, the phase current flows for almost the whole electrical cycle. Hence, there is a limited time for probing signals.

Methods that make use of the flux-linkage and current in an energized phase work better for high speed than for low speed. In particular, these methods suffer from the calculation of the flux-linkage at low speed and low torque levels. The reason is that the flux-linkage may result in large errors due to integration drift of the phase voltage.

The phase voltage can be either measured directly from each phase by isolation amplifiers or calculated by the switching state of the power converter. The former requires extra circuitry, in contrast, the latter does not and therefore it is simpler. However, it may be necessary to measure the d.c.-link to improve the calculation of the phase voltage. The calculation of the flux-linkage can be further improved if the phase resistance is changed as a function of temperature.

---

[13] These problems are amplified at low speed and low torque where the $R \cdot i$ drop is a significant part of the phase voltage.

It seems that the most suitable sensorless approach to cover the whole speed range with good accuracy should combine a method of probing pulses for low speed, and a method based on flux-linkage (observers) on an active phase for high speed.

An important factor that needs to be considered in any sensorless method is the startup sequence at standstill. Basically, there are three methods to initialize the rotor position.

1. Consists of energizing one of the phases in order to align the rotor to that phase. This method normally works. However, if the rotor is at the unaligned position, the rotor may not move even when the phase is excited.
2. Involves the excitation of two phases at a time with the same current level. In this case, the rotor will always be brought to a known overlap position.
3. Makes use of probing pulses that are injected in each of the phases. Then, the position can be estimated from the magnetization curves. However, if the specific initial rotor position is not required, the peak current of the probing pulses can be used to identify which phase needs to be energized first in order to start the motor in the desired direction.

For sensorless methods that do not estimate rotor position at standstill, the motor can be accelerated by applying a ramp of frequency in open loop (feedforward technique) as a stepper motor after identifying the starting phase.

Another important problem in sensorless control that has not been discussed in detail so far in the literature is to estimate the rotor position when the motor is already rotating. For instance, if the power to the motor is turned off for a short period of time and it is needed to turn it back on, or if the machine is being used as a generator. In these cases, it is necessary to estimate the position before any commutation. For this problem, the option is to inject probing pulses in all the phases and estimate the position from them, once the commutation is synchronized, the sensorless method can be swapped with other methods different from the probing pulses approach.

The type of load that is driven in an application defines the motor performance that is required, and thus the most suitable sensorless method. Figure 7.33 classifies the

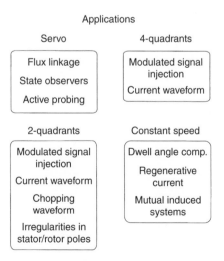

Fig. 7.33 Possible application of sensorless methods.

sensorless methods discussed in this chapter according to four main applications. The first one is servo applications, which require precise feedback of rotor position and speed of the motor over the whole speed range. In these applications an instantaneous rotor position is required. The most promising sensorless methods for servo applications could be: flux-linkage, state observers and active probing. However, a combination of these methods may be necessary in order to cover the required speed range.

On the other hand, there are applications where precise knowledge and high resolution of the rotor position are not of significant importance, but four-quadrant operation is still required. For this kind of application the most suitable sensorless methods could be: modulated signal injection and current waveform.

For applications that only require two-quadrant operation of the motor (i.e. the motor is driven in both directions but no braking is required), the above methods can be used with good performance, but other methods like chopping waveform and irregularities in stator/rotor poles may be applicable.

There are applications which require driving the load at constant speed with modest load torque transient capability. In these applications, sensorless methods like dwell angle compensation, regenerative current and mutual induced systems may be useful.

Methods such as flux-linkage and state observers can be suitable for applications where the cost is not an issue but robustness, high resolution and accuracy of the system in a harsh environment are important. It is evident that there is a trade-off between the resolution of rotor position estimation and the cost of the required extra components. In other words, there is no cheap solution for high resolution of rotor position estimation. However, the fact that every day there are more powerful digital signal processors in the market place with an ever dropping cost may open the window to high-resolution sensorless methods for most of the drive applications. These methods could potentially even be implemented in traditionally low-cost applications.

8

Torque ripple control in a practical application

Kevin McLaughlin
TRW Chassis Systems

8.1 General approach

The inherent qualities of the switched reluctance motor make it an ideal solution for an electrically assisted automotive steering gear. The rotor inertia is low, allowing precise, high-bandwidth steering control. There are no parts that wear other than bearings. The motor has no electrical failures that can cause unwanted torque when in an unpowered state. However, torque ripple and acoustic noise present some challenges (see, for example, Sahoo *et al.* (1999), Stephenson (1997) and Moreira (1992)).

Figure 8.1 shows the general arrangement of an electrically assisted steering gear. The gear is much like a classic rack-and-pinion assist gear except that the hydraulic assist function is replaced with a switched reluctance motor. The steering handwheel is mechanically connected to the pinion. Inside the pinion is a torque sensor that senses the torque applied by the driver. This torque signal is fed to the electronic control unit where it is processed to determine the desired motor torque. The motor rotor is mechanically connected to the rack through a ballnut/ballscrew mechanism which converts the motor torque to a rack force. The force generated on the rack turns the tyres to generate the desired steering effect.

Reducing torque ripple to imperceptible levels is critical in electrically assisted steering applications. Ripple is sensed as a torque variation at the steering handwheel as the motor turns – a phenomenon outside of all previous driving experiences. Hand-wheel torque variation is extremely objectionable and detectable by most drivers in the population. Similarly, acoustic noise generated by the motor must be low. Vehicle requirements usually specify that noise from the steering gear shall be imperceptible – a daunting task for a switched reluctance motor.

This chapter describes the approaches used to solve both the torque ripple and the acoustic noise problems, providing acceptable levels of steering performance. The torque ripple was eliminated by applying an adaptive approach, termed the 'black box' approach, in which knowledge of the internal physics of the motor was not critical. Acoustic noise was solved by an analytical approach which is described from first principles. The torque ripple solution is discussed first.

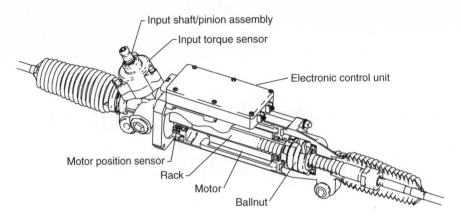

Fig. 8.1 Electrically assisted steering gear.

8.1.1 Black box approach

An 8/6 motor is used in the present example, i.e. a motor with 8 stator poles and 6 rotor poles. It is assumed that the electrical position of the motor rotor is sensed along with the current in the motor phases. This example deals specifically with controlling the motor torque ripple. It is assumed that the motor is already capable of generating torque substantially close to the d.c. level requested, but with substantial torque ripple.

To reduce torque ripple, the motor/controller set is viewed as a 'black box'. Figure 8.2 shows that the input to the black box is the commanded torque T_{cmd} and the output of the black box is the measured motor torque T_{meas}. Inside the black box are all the functions required to control and generate torque: current profile generation, current control, power amplifiers, current sensing, switched reluctance motor, etc. A scheme is applied to reduce the torque ripple to zero by simply manipulating the input to the black box. By 'zero torque ripple', it is meant that the mean of the measured torque at any rotor angle is equal to the torque command. No detailed internal knowledge of the motor is required; however, it is required that the relationship between input commanded torque and output measured torque be substantially predictable. Note that although the method is applied here to the switched reluctance motor, it is general and can be applied to any electrical machine.

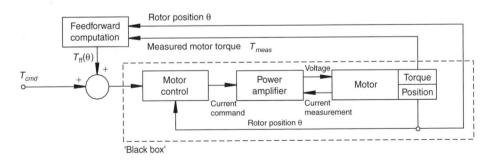

Fig. 8.2 'Black-box' approach.

8.1.2   Previous approaches

Several approaches to controlling torque ripple exist (see Chapter 6 and Krause (1986)). Generally, current profile tables are defined for all motor phases. The tables may contain reference values for the phase currents as a function of rotor position, or coefficients of equations which relate the reference currents to the rotor position. They are developed in such a way that if the current regulator forces the phase currents to follow the reference profiles, the motor torque will remain substantially constant as the motor rotates through an electrical cycle. Motor position is measured and the corresponding phase current commands are generated either through a look-up table or an equation. The current in the phase is measured and controlled or regulated to follow the current profile. Typical profile tables for the example motor are shown in Figure 8.3.

The main problem with this approach is that *complete knowledge of how the SR motor generates torque must be known prior to generating the tables*. Successful ripple control requires knowledge of the motor torque generated not only with single-phase excitation, but also with dual-phase excitation.

The motor torque generated with single-phase excitation is nonlinear, but easily measured. Figure 8.4 shows a set of measured 'static torque curves', indicating the motor torque as a function of electrical angle for a single-phase current of 10, 20, ... , 60 A. Current profile tables for single-phase operation can be generated from the static torque curves (see Chapter 6).

The prediction of motor torque is not so simple when more than one phase is excited. Mutual effects between phases can become dominant, especially in machines designed with tight size constraints, so it cannot be assumed that the torque contributions of the two phases add linearly: i.e. the superposition assumption becomes invalid. If $T_A(i_A, \theta)$ is the torque measured at rotor position θ when phase A is singly excited with current

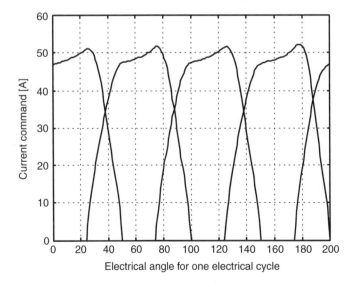

Fig. 8.3 Graph of typical current profile tables.

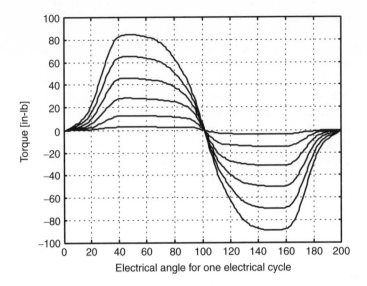

Fig. 8.4 Static torque curves (single-phase excitation) at current levels of 10–60 A.

i_A, and similarly $T_B(i_B, \theta)$ for phase B, then

$$T(i_A, i_B, \theta) \neq T_A(i_A, \theta) + T_B(i_B, \theta) \tag{8.1}$$

i.e. the total torque $T(i_A, i_B, \theta)$ generated when both phases are excited is not equal to the sum of the two individual phase torques.

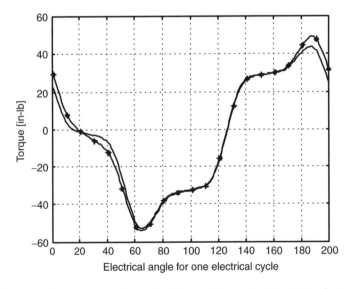

Fig. 8.5 Dual-phase excitation: superposition (asterisk) vs. simultaneous (solid) excitation for 30 A in phases A and D.

This inequality of (8.1) is illustrated in Figure 8.5. The solid line indicates the right-hand side of (8.1) for 30 A in both phases A and D (the superposition assumption): phase A was excited with 30 A and the torque/angle curve measured, phase B was excited with 30 A and the torque/angle curve measured, and then the two curves were summed algebraically. The line with asterisks represents the measured output torque with 30 A simultaneously in phases A and D; that is, the left-hand side of (8.1). For the 8/6 motor, simultaneous excitation of phases A and D creates the longest flux paths in the stator yoke and is therefore the worst case for superposition. Unless some estimate of the mutual effects is found, torque ripple will exist if superposition is assumed valid.

Many other practical issues can impact torque ripple and invalidate the superposition assumption: position measurement accuracy, current measurement accuracy, current control, initial current profile generation, computational delays, etc. These parasitic effects are difficult to model, yet they must be taken into account to achieve low torque ripple.

8.1.3   The black box approach: problem definition

The black box approach 'incorporates' the mutual and parasitic effects by directly exploiting the input/output behaviour of the motor/controller system as a 'black box'. The basic requirement is to control the motor such that the measured motor torque T_{meas} is identically equal to the desired motor torque T_{cmd} at all motor angles. Assuming that the motor torque and rotor angle are measured, this requirement can be met by viewing the motor/controller as a black box and solving a single problem: *find a feedforward torque $T_{ff}(\theta)$ that when added to the torque command T_{cmd} makes the motor output torque T_{meas} identically equal to the commanded torque at all angles.* Mathematically,

$$T_{req}(\theta) = T_{cmd} + T_{ff}(\theta) \qquad (8.2)$$

where $T_{req}(\theta)$ is the torque requested of the motor such that it has zero ripple. A block diagram to achieve this is shown in Figure 8.6. Effectively, the torque requested of the motor is modulated as a function of electrical angle to cancel out torque ripple. Ideally, the feedforward torque $T_{ff}(\theta)$ is 180° out of phase with the motor torque ripple. However, owing to nonlinear effects, this will not be strictly true.

Though the basic concept of the black box approach is simple, practical implementation may be complicated. Motor torque ripple must be controlled at all torque levels and in all four operating quadrants (i.e. positive and negative torque; positive and negative speed). Torque measurements contain noise and test-stand dynamic effects that must be rejected. Inaccuracies exist in other measurements and in the control system. All these issues must be addressed.

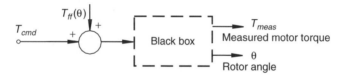

Fig. 8.6 Find T_{ff} such that $T_{meas} = T_{cmd}$.

The following three sections describe the analytical tools required to implement the torque ripple reduction algorithm. Each section stands alone and the final algorithm is described in Section 8.5. Measurement results at a single torque level are presented in Section 8.6 and the extrapolation of the algorithm to all torque levels is contained in Section 8.7.

8.2 The zero phase or anti-causal filter

Given a set of measured data, it is possible to filter the data such that zero phase error exists between the measured and the filtered data. This is critical when using the black box approach because the feedforward torque will cancel the motor torque ripple only if it is applied 180° out of phase with the motor torque ripple. If the phase of the feedforward torque $T_{ff}(\theta)$ is not sufficient to cancel the torque ripple, many iterations will be required to achieve low ripple. If the phase error is too great, the algorithm may not converge and it will then fail to reduce the motor torque ripple to an acceptable level. In general the torque-smoothing algorithm is more robust to gain errors than to phase errors, so every attempt is made to minimize the phase error.

In any torque measuring system, the measured torque is filtered to reduce noise and unwanted high-frequency content. In the present example, the torque signal is sampled at 3 kHz with an analogue-to-digital converter (ADC) and a digital filter is applied to the raw torque signal to remove unwanted high-frequency data. This filter induces phase lag between the measured and filtered data and it is specifically this phase lag that must be recovered.

Additional phase lags between the motor torque command and the measured torque are induced by anti-aliasing filters, mechanical dynamics, sampling effects, etc. These effects are often small and initially are neglected, so only the phase of the digital torque filter is compensated. It is later shown how to compensate the additional phase lags.

Assume that the measurements exist as a sequence of numbers equally spaced in time. It is desired to remove all frequency content above some frequency ω_b without imparting a phase lag with respect to the measured data. One method is to filter the data through a digital filter with a break frequency of ω_b, then 'reverse time' and feed the filtered signal back through the same filter. The resulting twice-filtered signal is reversed in time again and the final result has twice the gain reduction of the original filter without any phase lag. Because the data is reverse-time filtered, the process is noncausal and the entire filtering procedure is referred to as an 'anti-causal' filter.

As an example, consider a time domain signal $u(t)$:

$$u(t) = \sin(\omega t). \tag{8.3}$$

If $u(t)$ is passed through a filter with gain of K and phase of φ at frequency ω, then the output of the filter $y(t)$ is

$$y(t) = K \sin(\omega t + \varphi). \tag{8.4}$$

Time is reversed by substituting $t = -\tau$ into (8.2):

$$y(\tau) = K \sin(-\omega \tau + \varphi) = -K \sin(\omega \tau - \varphi). \tag{8.5}$$

If the signal $y(\tau)$ is filtered through the same filter used in (8.1), the result is:

$$y'(\tau) = -K^2 \sin(\omega\tau - \varphi + \varphi) = -K^2 \sin(\omega\tau). \tag{8.6}$$

Substituting $t = -\tau$ into (8.5) yields

$$y'(t) = K^2 \sin(\omega t). \tag{8.7}$$

The signal $y'(t)$ has twice the gain reduction of the original filter with zero phase error.

Consider a data stream sampled at 200 Hz for 1 second (a sample of 200 points in 1 second). Let $t = 0, 0.005, 0.01, \ldots, 1.0$, and suppose that the data $y(t)$ is defined by

$$y(t) = 1.4 \sin((3 \times 2\pi)t + 1) + 0.7 \sin((6 \times 2\pi)t + 2.4)$$
$$+ 0.3 \sin((9 \times 2\pi)t + 0.4) + \text{rand}(200, 1) \tag{8.8}$$

where rand(200,1) is the Matlab^{TM} random number generator that generates a pseudo-random sequence of 200 numbers with a uniform distribution between 0.0 and 1.0. Figure 8.7 shows a plot of the noisy signal $y(t)$ and the much smoother anti-causally filtered signal. The Matlab^{TM} routine 'anti.m' shown in Figure 8.8 was used to compute the anti-causally filtered signal. The digital filter used in both the forward and reverse time is first-order with a break frequency at 16 Hz.

The routine 'anti.m' uses a first-order filter. Higher-order Butterworth filters are generally used to achieve more uniform gain characteristics below the break frequency and steeper fall-off above the break frequency. The torque ripple reduction algorithm uses a sixth-order Butterworth filter.

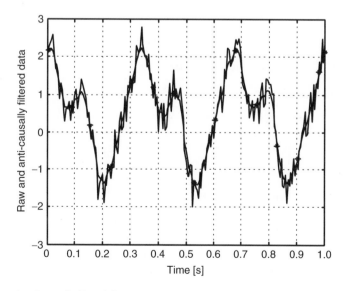

Fig. 8.7 Raw and anti-causally filtered data.

```
function [uhat]=anti(u,pole,ndim)
%****************************************************************
%  This routine anti-causally filters a stream of data through
%  a first order filter with z="pole".  The data is assumed
%  to be in a vector of length 200 points and is also assumed
%  to repeat the same pattern after the 200th point.
%
%  inputs:  u    = input vector of 200 points
%           ndim = some multiple of 200 so that
%                  no transients occur in the
%                  filtered output
%           pole = z-domain location of the pole.
%                  z=exp(s*2*pi/200) where s  is
%                  defined such that s=1 is for
%                  1 cycle every 200 points, s=2
%                  is for 2 cycles every 200 points, etc.
%                                      K. McLaughlin
%                                      January 17, 1994
%****************************************************************
%  yk   = forward time filtered signal.
%  ykk  = reverse time filtered signal.
%  unew = input signal repeated ndim/200 times.
%****************************************************************
yk = [1:1:ndim]'*0.0;
ykk=[1:1:ndim]'*0.0;
%****************************************************************
%  Assume that the input vector u is repeating.  Make the vector
%  unew be the vector u repeated 6 times.  This will avoid
%  transients at the beginning and end of the sequence.
%****************************************************************
unew = [u' u' u' u' u']';
%****************************************************************
%  forward time filter the input
%****************************************************************
for i=2:ndim,
   yk(i) = unew(i) - pole*( unew(i) - yk(i-1) );
end;
%****************************************************************
%  Take the last 200 points of yk. Create another vector
%  "uunew" with this filtered data.  Repeat as above with u.
%****************************************************************
yktmp = yk(ndim-199:1:ndim);
uunew = [yktmp' yktmp' yktmp' yktmp' yktmp']';
%****************************************************************
%  reverse time filter the data in "uunew".  Note that the
%  filter starts at the last point in uunew and ends at the
%  first point in the data.
%****************************************************************
for i=ndim-1:-1:1,
  ykk(i) = uunew(i) - pole*(uunew(i) - ykk(i+1) );
end;
%****************************************************************
%  The anti-casually filtered data is the last 200 points of
%  the reverse time filtered data.
%****************************************************************
uhat = ykk(1:1:200);
```

Fig. 8.8 Anti-causal filter code.

8.2.1   Anti-causal filters and torque ripple data

When applying anti-causal filtering to motor torque ripple data, recognize that torque ripple can be viewed either in the time domain or the spatial domain. In the time domain, torque is plotted against time; in the spatial domain, torque is plotted against electrical angle. For an 8/6 switched reluctance motor, an electrical cycle covers $60°$ (mech.) If the phases are energized A, B, C, D, A,... sequentially, the motor rotates $60°$ (mech.), that is, $360°$ (elec.).

The spatial domain is convenient because physical motor and control phenomena occur at spatial frequencies that are invariant with respect to motor speed. For example, the 8/6 motor commutates four times per electrical cycle, corresponding to a spatial frequency of $N = 4$ cycles per electrical cycle. This spatial frequency is invariant; i.e. it is independent of the motor speed.

For a motor rotating at a constant speed, there is a one-to-one relationship between time-domain frequencies ω and spatial frequencies N. For an 8/6 motor, the relationship is computed as:

$$W_{rpm} \left[\frac{rev}{min} \right] \times \left[\frac{6 \text{ elec. cycles}}{rev} \right] \times \left[\frac{1 \text{ min}}{60 \text{ sec}} \right] \times N \left[\frac{cycle}{\text{elec. cycle}} \right]$$

$$= \omega \left[\frac{cycles}{sec} \right] = f(Hz) \tag{8.9}$$

or

$$\frac{W_{rpm}}{10} \times N = f \tag{8.10}$$

where W_{rpm} is the motor rotational speed in revolutions per minute, N is the spatial frequency in cycles per electrical cycle, and f is the time-domain frequency in Hz. (8.10) is useful when recovering the phase from filtered motor data.

In order to remove unwanted high-frequency information from the torque ripple data, two techniques are proposed: a spatial/spatial filter and a time/spatial filter.

In the *spatial/spatial filter*, the raw measured torque data is averaged as a function of electrical angle. Given a torque measurement $T(n\tau)$, where T is the torque, τ is the sample interval, and n is the nth sample, the measured torque is averaged as a function of motor electrical angle θ:

$$T_{avg}(j) = \frac{1}{N(j)} \sum_{n=1}^{m} T(n\tau), \quad (j-0.5) \le \theta_j \le (j+0.5), \quad j = 1, 2, \dots, 200.$$
$$\tag{8.11}$$

In (8.11) there are $N(j)$ samples at each electrical angle. The vector T_{avg} in this case has length 200, meaning that there are 200 electrical angles per electrical cycle. This number is arbitrary and could easily be changed to 100 or 360 or whatever is convenient. The averaged torque vector T_{avg} is anti-causally filtered to remove any high-frequency content using the Matlab^{TM} routine 'anti.m'. Because the vector of (8.11) is already a spatial vector, the break frequency for the anti-causal filter must be specified as a spatial frequency. It has been found that by filtering with a break frequency of $N = 16$ cycles/electrical cycles, good convergence of the torque ripple reduction algorithm occurs.

The spatial/spatial filter is easy to implement but has one drawback: aliased data can creep into the averaged torque signal. For example, suppose that the motor torque ripple data is sampled at 3 kHz and that the motor is rotating at 20 rev/min. Because there are only 200 points in the vector, the maximum spatial frequency that can be realized, i.e. the Nyquist frequency, is $N = 100$ cycles per electrical cycle. From (8.10), this corresponds to a time-domain frequency of $f = 200$ Hz. However, because the raw data from the analogue-to-digital converter is sampled at 3 kHz, it may contain frequency information up to 1500 Hz. If the original measured torque data has a frequency content above 200 Hz, this frequency content will be aliased into the average torque vector. As long as the torque ripple data has no high-frequency content, the spatial/spatial filter is sufficient. If the high-frequency content is unknown, then it is recommended to use the time/spatial anti-causal filter.

In the *time–spatial filter*, the data is first filtered in the time domain. The filtered torque ripple is averaged using (8.11). After averaging, the phase is recovered in the spatial domain by creating a spatial filter with the same break frequency as the time-domain filter, using (8.10). So, for example, suppose the motor is rotating at 20 rev/min and a time-domain filter with a break frequency of 30 Hz is applied to the measured data. From (8.10), this corresponds to a spatial frequency of 15 Hz. Using (8.11) to average the filtered torque ripple data over electrical angles, the resulting spatial vector is passed only through the reverse time portion of the anti-causal filter; the phase lost by time domain filtering the raw torque signal is recovered in the spatial domain.

The advantage of the time/spatial algorithm is that all high-frequency data is removed from the torque signal prior to averaging. This can be likened to adding an anti-aliasing filter in the time domain before averaging the torque ripple at discrete electrical angles. The effect on torque ripple reduction is better control of the frequencies included in the torque ripple estimate, and reduced likelihood of high-frequency data above the spatial Nyquist frequency aliasing into the torque ripple estimate.

In typical applications, a sixth-order Butterworth filter is used to filter the measured torque ripple data in the time domain. Since twice the gain reduction with zero phase error occurs because of anti-causal filtering, a roll-off of 240 dB/decade occurs in the anti-causally filtered data.

8.2.2 Compensated miscellaneous dynamics

The time/spatial filter compensates only the phase errors induced by the digital filter applied to the raw torque sensor signal. However, the raw torque sensor signal contains many other phase and gain errors with respect to the actual commanded motor torque: sampling effects from the analogue-to-digital conversion, dynamic effects in the test stand, analogue anti-aliasing filters, etc. Rather than attempting to quantify each of these effects analytically, it is best to measure the frequency-domain transfer functions between the motor torque command and the raw measured torque:

$$G(\omega) = \frac{T_{meas}}{T_{cmd}} \tag{8.12}$$

where T_{meas} is the raw measured torque from the analogue-to-digital converter and T_{cmd} is the motor torque command. Many methods exist to find $G(\omega)$; however, a sine sweep on the torque command is typically the most accurate. One trick in obtaining a repeatable measurement is to have the motor rotating and applying an offset torque when measuring $G(\omega)$. By calculating $G(\omega)$ as defined in (8.12), a measure of all the miscellaneous dynamic effects that require compensation is achieved.

If the torque data is sampled at a very high rate and the test-stand dynamics are such that no resonances occur near the desired break frequency, then the transfer function $G(\omega)$ will have essentially unity gain and zero phase below the break frequency and can be neglected. However, if $G(\omega)$ does not have unity gain and/or zero phase below the break frequency, then it must be compensated for. The black box approach requires knowledge of how the torque command must be modified to achieve zero measured torque ripple, and (8.12) represents the physical transfer function between these two quantities. Only the phase error due to $G(\omega)$ need be compensated in the anti-causal portion of the filter to assure convergence.

To achieve this, the time/spatial filter is modified. As before, the measured torque signal is filtered through a digital filter and averaged. The averaged signal contains gain and phase errors due to both the digital torque filter and the transfer function $G(\omega)$. The spatial portion of the anti-causal filter is modified to contain not only the digital filter, but also the transfer function $G(\omega)$. This allows compensation of phase errors due to test-stand dynamics, sampling, etc. No gain errors are compensated for; however, it has been found that phase is the critical parameter requiring compensation. So long as the feedforward vector is lined up properly out of phase with the motor torque ripple, it will necessarily converge to zero torque ripple while compensating for gain errors. If both gain and phase are incorrect, the process may never converge.

In general, the test stand and data handling system should be designed such that the transfer function $G(\omega)$ has unity gain and zero phase in the frequency range in which torque ripple is critical. If so, then this transfer function can be neglected and only the phase errors induced by the digital filter need be compensated.

8.3 Spatial frequency-domain analysis

The concept of spatial frequency-domain analysis and its application to torque ripple reduction are explained in this section. Though the material presented is not critical to understanding the ripple reduction algorithm, it is useful in understanding the final results and the strategies chosen to reduce the torque ripple.

By applying Parseval's theorem to the measured torque data, the torque ripple contribution within any frequency range is easily viewed graphically. Parseval's theorem states that the variance of a time-domain signal is equal to the integral of the power spectral density (PSD) over all frequencies,

$$\sigma^2 = \frac{1}{N} \sum_{n=1}^{N} x(n\tau)^2 = \frac{1}{2\pi} \int_{-\infty}^{+\infty} P(\omega)\,d\omega. \tag{8.13}$$

If the time sequence is real, then the PSD is symmetric with respect to the zero frequency point and (8.13) can be modified as

$$\sigma^2 = \frac{1}{N} \sum_{n=1}^{N} x(n\tau)^2 = \frac{1}{\pi} \int_{0}^{+\infty} P(\omega)\,d\omega. \tag{8.14}$$

By plotting the integral of the PSD as a function of frequency, the energy contribution within each frequency band is clearly seen. If the integral of the PSD is plotted with respect to spatial frequency rather than time-domain frequency, the frequencies become invariant with respect to motor speed. Since spatial frequencies relate directly to motor control and motor construction, the torque ripple due to these phenomena are detectable graphically.

An example of a spatial frequency related to motor control is the commutation of the phase currents. In the 8/6 motor, there are four commutations per electrical cycle, represented by the spatial frequency $N = 4$. Harmonics of this frequency also appear in torque ripple data, i.e. $N = 8$, 12, 16, etc.

An example of a spatial frequency related to motor construction is torque ripple due to rotor variation. Designate the three sets of rotor poles as x, y, and z and define a 'rotor electrical cycle' as the electrical angle over which the sequence xyz repeats. For the 8/6 motor, a rotor electrical cycle is $270°$ (elec.). Suppose the y set of rotor poles in the 8/6 motor is ground to a different diameter from that of the x and z sets. The effective airgap between the stator poles and the y rotor poles will be different from the effective airgap between the stator poles and the x and z rotor poles. Therefore, the torque will be different when the y pole is 'pulled in', as compared with the x and z poles. Since the rotor cycle repeats every $270°$ (elec.), the torque variation due to the rotor variation will occur every 360/270 or $N = 1.333$ cycles per electrical cycle.

Noninteger frequencies like the rotor electrical cycle are critical because their content defines the lower bound on ripple reduction. Recall that zero torque ripple is defined such that the mean of the torque over an electrical angle is equal to the commanded torque. If the torque ripple contains noninteger frequencies, then at any given electrical angle these frequencies will average out to zero over an entire revolution of the motor and not affect the mean. They will not be compensated for in the ripple reduction algorithm. Other examples of noninteger spatial frequencies are motor bearing friction and torque measurement misalignments that occur once and twice per motor revolution (1/6 and 1/3 per electrical cycle). Since these again average to zero at any electrical angle over a motor revolution, they will not be compensated for in the ripple reduction algorithm.

Since the torque ripple algorithm only compensates for integer values of N, the noninteger terms define the lowest level to which ripple can be reduced by compensation in the electrical cycle frame. However, if rotor variation or friction variation are unavoidable issues, then defining the feedforward torque as a function of mechanical angle rather than electrical angle will allow them to be compensated.

8.3.1   An example of spatial frequency-domain analysis

Figure 8.9 shows motor output torque ripple for a 50 in-lb command. The motor is rotating at a constant 20 rev/min and the peak-to-peak torque ripple is about 7 in-lb. Figure 8.10 shows the same data plotted against electrical angle. The tables from Figure 8.2 were used to command the motor. The bulk of the torque ripple occurs during the AD phase transition. Though torque spikes are clear in the time-domain data, the cause of the spikes is not clear. By viewing the torque ripple in the spatial domain, the fact that the AD transition is the main problem is evident. Frequency-domain analysis is similarly useful when viewed spatially rather than temporally.

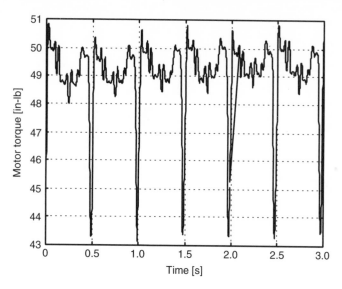

Fig. 8.9 Uncompensated torque ripple: motor torque vs. time.

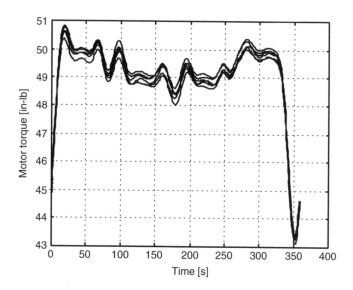

Fig. 8.10 Uncompensated torque ripple plotted spatially, i.e. against electrical angle.

Figure 8.11 shows the PSD of the time-domain data. The data contains many spikes, but again, it is not directly evident which motor phenomena are causing the spikes. Figure 8.12 shows an integral of the PSD plotted against spatial frequency; the PSD from Figure 8.11 was integrated using (8.14) and the results plotted against spatial frequency. It is seen that torque ripple occurs at $N = 4, 6, 8, 10, \ldots,$ cycles/electrical cycle.

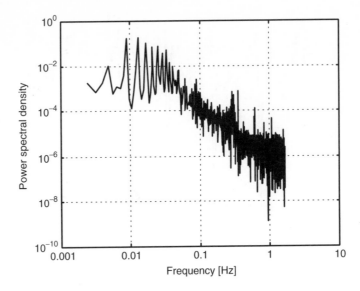

Fig. 8.11 Power spectral density of torque ripple: PSD for base tables without T_{ff} correction.

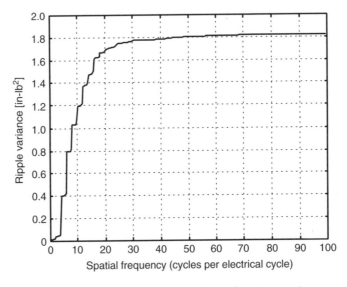

Fig. 8.12 Integral of PSD (power spectral density) for base tables without T_{ff} correction.

8.3.2   Selection of the anti-causal filter frequency

The anti-causal filter frequency determines the frequency above which no motor torque ripple will be compensated. One method to select this frequency is to rotate the motor at a predetermined speed and measure the torque ripple for a base set of tables. Compute the power spectral density of the torque ripple and integrate using (8.14) to obtain the total variance of the measured ripple, σ_{total}^2. Determine the amount of torque ripple in terms of a variance that is acceptable, $\sigma^2_{acceptable}$, and assuming that the ripple is uncorrelated, compute the desired amount of reduction as:

$$\sigma^2_{reduction} = \sigma^2_{total} - \sigma^2_{acceptable}. \tag{8.15}$$

Find the frequency at which the variance of the ripple is equal to $\sigma^2_{reduction}$; this is the frequency below which all torque ripple must be reduced. Because the torque ripple is generally not uncorrelated with electrical angle, (8.15) is a weak approximation at best. However, it is a good place to start.

In the previous example, the total variance of the torque ripple is 1.85 (in-lb)^2. Suppose it is desired to reduce the torque ripple by 90% or 1.66 (in-lb)^2. From Figure (8.12) it is seen that a variance of 1.66 corresponds to a frequency of $N = 16$. So if all ripple is reduced to zero below $N = 16$, the motor will have a resulting ripple with variance of about 0.2 (in-lb)^2.

8.4   Initial profile table generation

8.4.1   Two-dimensional interpolation and table storage

It is desired to obtain smooth torque control not only as the motor rotates but also at changing torque levels. Because processor memory is finite, the current profile tables are stored at discrete angle and torque levels. Storing coefficients for equation-based profiles is an option but the equations must be complex enough to capture sufficient frequency content for zero torque ripple. In the present application the profile tables are at increments of 10% of maximum torque; each profile table contains the current commands for phases A, B, C and D at 200 discrete angular increments within the electrical cycle.

In order to obtain smooth and quiet operation of the motor, a two-dimensional interpolation is used to compute current commands (McLaughlin and Gluch, 1995). For example, suppose the motor is at $\theta = 35.4°$ (elec.) and the torque command is 57% of the maximum torque. The algorithm looks up the current command for the four phases in the 50% table at $\theta = 35°$ (elec.) and $\theta = 36°$ (elec.) and in the 60% table at the same angles. Then a two-dimensional linear interpolation is performed to compute the currents at the 57% torque command and 35.4° (elec.)

The two-dimensional interpolation keeps the current commands piecewise-continuous over all command conditions. This keeps the acoustic noise low and allows the motor to achieve low levels of torque ripple between the torque levels of the defined profile tables.

In the steering gear application, smooth torque ripple is required in all four operating quadrants. It has been found that the profile tables generated in Q1 do not produce low torque ripple when used to command the motor in Q2. In fact, the zero ripple tables are different in Q1 and Q2, and similarly for Q3 and Q4. Schemes do exist to use a single set of profile tables in all four operating quadrants, e.g. US Patent No. 5,998,952; however, low torque ripple may drive the design to a separate set of profile tables in all four operating quadrants.

8.5 Torque ripple reduction algorithm

A block diagram of the ripple reduction algorithm is shown in Figure 8.13. The algorithm as shown will compute the feedforward torque $T_{ff}(\theta)$ to achieve zero torque ripple at a single torque command and speed. A time/spatial anti-causal filter is utilized (McLaughlin and Parker, 1999).

The process begins by commanding the motor to a constant speed and constant torque command. It is assumed that a set of base tables has been loaded into the controller so that the mean output torque is 'close' to the desired command. 'Close' means that the peak-to-peak torque ripple is generally less than 20% of the maximum rated motor torque. Actually, so long as the local gain between the commanded torque and the measured torque is greater than zero and less than two at all angles, the algorithm will always converge. For example, if the motor is commanded to 50 in-lb and the measured torque ripple designated as $T_{50}(\theta)$, and then the motor is commanded to 51 in-lb and the corresponding measured torque designated as $T_{51}(\theta)$, then the algorithms will converge so long as $0 < (T_{51}(\theta) - T_{50}(\theta)) < 2$ for all θ. This condition is difficult to violate.

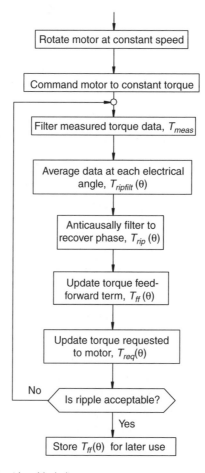

Fig. 8.13 Ripple reduction algorithm, block diagram.

The torque and the electrical angle data are collected for one revolution of the motor. As the torque data is collected, it is digitally filtered and averaged at each electrical angle. In the present example, the motor is rotated at 20 rev/min, the torque command is 50 in-lb, and the digital filter is a Butterworth filter with a break frequency of 32 Hz. Designating the filtered torque ripple averaged at each angle as $T_{meas}(\theta)$, compute the filtered ripple $T_{ripfilt}(\theta)$ with respect to the command as

$$T_{ripfilt}(\theta) = T_{cmd} - T_{meas}(\theta). \tag{8.16}$$

Next, the phase lost in the time-domain digital filter is recovered using the spatial anti-causal filter. At 20 rev/min, the time-domain frequency of 32 Hz corresponds to a spatial frequency of $N = 16$ cycles/electrical cycle. Denote the anti-causal filtered ripple estimate as $T_{rip}(\theta)$. Next, the feedforward torque is updated:

$$T_{ff}(\theta) = T_{ff}(\theta) + KT_{rip}(\theta). \tag{8.17}$$

Initially, the vector $T_{ff}(\theta), \theta = 1, 2, \ldots, 200$ is set to zero so that, the first time through the algorithm, T_{ff} is equal to KT_{rip}. The gain K is used in the iteration to account for the fact that the local gain between commanded and measured torques may not be unity at all electrical angles. In the present example, $K = 1.0$.

Next, the torque request to the motor is updated to

$$T_{req}(\theta) = T_{cmd} + T_{ff}(\theta) \tag{8.18}$$

and the whole process starts again. If the torque ripple is acceptable, then the algorithm is terminated and $T_{ff}(\theta)$ stored for later use. If the measured ripple is unacceptable, the process goes back to the point at which the torque is measured and repeats.

The algorithm is structured such that the torque ripple measurements and feedforward torque updates are performed during alternate motor revolutions. For the example of the motor rotating at 20 rev/min, torque is measured, filtered and averaged during the first revolution. The processor then has one revolution or 3 seconds to filter the data anti-causally and update the feedforward torque estimate. During the third revolution, motor torque ripple is again collected, filtered and averaged. During the fourth revolution, the feedforward torque is again updated. This process continues until the ripple is at an acceptable level. Generally, two to four iterations are required to achieve acceptable torque ripple. At 20 rev/min, this corresponds to a maximum of 8 revolutions of the motor (2 revolutions per iteration) or 24 seconds.

Because the motor is running continuously during this process, minimizing the time to complete the ripple reduction is critical to keep the motor temperature low. The closer the base tables are to the desired zero ripple, and the better the input/output gain map is within the tables, the quicker the algorithm will converge.

8.6 Experimental results at a single torque level

The torque ripple reduction algorithm was applied to the 8/6 switched reluctance motor. The uncompensated data was shown in Section 8.3. Figure 8.14 shows the torque ripple after applying the algorithm of Section 8.5. The peak-to-peak torque ripple was reduced to about 1.0 in-lb. Figure 8.15 shows the torque ripple plotted versus electrical angle.

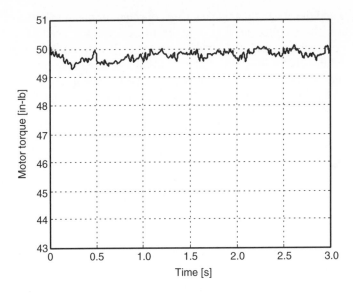

Fig. 8.14 Motor output torque vs. time with feedforward torque.

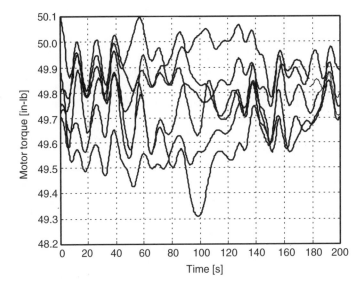

Fig. 8.15 Ripple torque for one electrical cycle with feedforward torque.

Note that at each electrical angle, the torque ripple averages to about the same values. The highs and lows at each electrical angle are due to the variation in torque at different mechanical positions for the same electrical angle. Figure 8.16 shows the PSD for the uncompensated and compensated cases. The anti-causal pole was set to 16 cycles per electrical cycle and the compensated ripple has nearly zero energy below the break frequency of the filter.

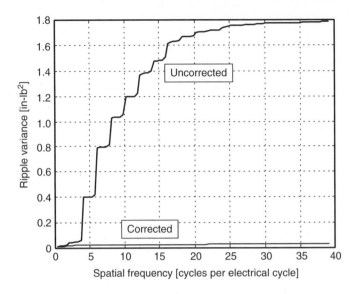

Fig. 8.16 Integral of PSD for base tables with and without T_{ff} correction.

8.7   Ripple reduction at all torque levels

Once the feedforward torque vector can be computed at a single torque level, it is a simple process to make the motor smooth over all torque levels and in all operating quadrants. As stated earlier, the profile tables are stored at increments of 10% of maximum torque. In fact, motors have been run in which the profiles have been stored at 20% of maximum torque, reducing the load on memory. The torque level discretization is somewhat arbitrary and determined by how well the motor performs between torque levels. Applying the two-dimensional interpolation will improve performance.

For this example, assume that the motor profile tables are stored at increments of 10% of maximum torque. Assume also that there are 200 angular points at which the currents in the four phases are specified. The steps to compute zero torque ripple profile tables for the motor are as follows:

1. Compute the feedforward vectors at each torque level and in each operating quadrant, and store in memory. For the present example, this means 40 feedforward vectors of 200 points each.
2. Disable the motor current controller so that the motor phases are not energized.
3. Set the operating quadrant to Q1 (positive speed, positive torque).
4. Set the motor command to 10% of full torque.
5. At each motor angle from 1 to 200, compute the phase current commands for the torque command, this being the sum of the percentage commanded and the corresponding feedforward torque for the zero ripple torque command. Store the zero ripple phase currents in memory.
6. Increment the torque requested by 10%. If the torque is less than or equal to 100%, go to step 5 and continue. If the torque is greater than 100%, go to step 7.

7. Increment the quadrant flag and return to step 3. If all four quadrants have been completed, exit.

The current commands stored in memory in step 5 above will produce a zero torque ripple level when used to command the motor. They consist of the d.c. torque command plus the feedforward torque command.

8.8   Acoustic noise

Though the mechanisms of acoustic noise generation are generally well known (Cameron and Lang, 1992; Wu and Pollock, 1995), models that describe them are not readily available. A derivation is presented that provides a simple model of acoustic noise generation in the switched reluctance motor. The model is general and may have application in other electric machines. It was developed utilizing concepts contained in the modelling technique known as 'bond graphs' (Karnopp, 1975; Paynter, 1977). Bond graphs were developed to model mixed dynamic systems: electrical, mechanical, thermal, fluid, hydraulic, etc. The essence of bond graphs is that power can flow in two directions between energy domains, and this power is conserved. For example, in a d.c. brush motor, the torque T is related to the current i by the torque constant k_T,

$$T = k_T i. \tag{8.19}$$

Similarly, the back-EMF e is related to the angular velocity ω through the same constant,

$$e = k_T \omega. \tag{8.20}$$

Solving for k_T in (8.20) and substituting in (8.19) yields the power balance relationship:

$$T\omega = ei. \tag{8.21}$$

Both sides of the equation represent power: the left-hand side represents the mechanical power and the right-hand side the electromagnetic power. The term 'power variables' refer to appropriate variables which when multiplied together yield power. For example, consistent power variables are: linear force and velocity; current and voltage; angular velocity and torque; temperature and rate of change of entropy.

The concept of power variables was applied to the generation of noise in the switched reluctance motor. It is known that shell vibration causes acoustic noise and that the shell deflection is induced by phase current in attracting the stator shell towards the rotor poles. The force is assumed to be proportional to the current,

$$F = k_1 i. \tag{8.22}$$

In the mechanical domain, the power variables are force and velocity. In the electromagnetic domain, the power variables are current and voltage. Assuming that force is proportional to current, then for power to be conserved, there must be an induced voltage e proportional to the velocity of the shell,

$$e = k_1 \frac{dx}{dt} \tag{8.23}$$

where dx/dt is the velocity of the shell with respect to the rotor. So as the force moves the shell towards the airgap, an EMF e is induced across the coil. This voltage is present only if the shell has a nonzero velocity. This implies that the motor acts equally effectively as a speaker or a microphone.

Figure 8.17 demonstrates this effect. The 8/6 motor tested was used for high-voltage applications and had a coil winding resistance of about 6 ohm. A 12 V battery was connected across one of the phase windings, driving a current of 2 A in the motor. The outside of the shell was struck with a hammer and the current waveforms shown in Figure 8.17 were measured. Figure 8.18 is a time expansion of Figure 8.17 to illustrate more clearly the frequency of the current oscillation. It is clear that the hammer excited a radial–axial vibration mode of the stator and that the motor acted like a microphone. The oscillation is nearly a perfect impulse response for a second-order system, showing that the shell has a natural frequency of 1934 Hz. The damping is estimated to be 5.4% from the data.

Figure 8.20 shows a block diagram representing both the current/force and the voltage/velocity relationships. The coil is no longer just a simple first-order L-R system, since it includes the structural coupling of the shell vibration to the induced EMF. The following section provides an electrodynamic derivation of this model from the equations of the second-order system shown in Figure 8.19 (Krause, 1986). The structural model for the motor is typical of the type used to model bending modes; the gain of the bending mode represents the amount of deflection for a nominal input force. Note that in general, an infinite number of modes are present in any structure and these can be modelled as linear sums to the voltage. Also, each of the modes has a different transfer function to acoustic noise, and this effect can also be included using the general structure of Figure 8.19 by assuming linear superposition of flexible motion – a valid assumption for the first few vibration modes of the stator.

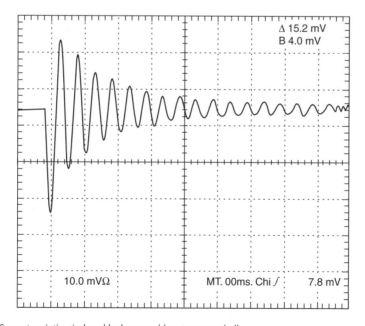

Fig. 8.17 Current variation induced by hammer blow to stator shell.

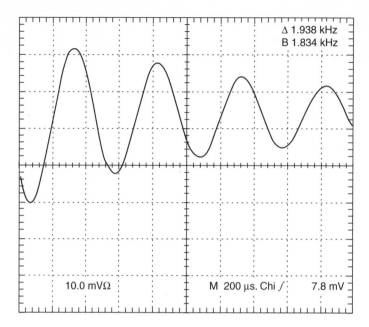

Fig. 8.18 Expanded waveform from Fig. 8.17.

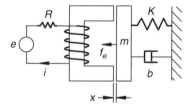

Fig. 8.19 The microphone model; block diagram.

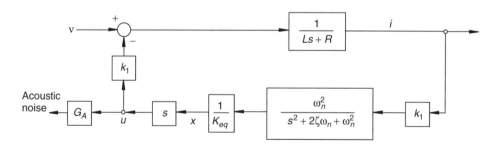

Fig. 8.20 Microphone model of linearized electromagnet.

The validity of this model is further illustrated by the deadbeat control technique proposed by Pollock (1995). Pollock showed via test that by stepping the voltage command to the coil at equal steps separated in time by half the natural frequency of the shell, that the acoustic noise is reduced. This technique is applied to the derived model to show that shell vibration is greatly reduced.

8.8.1 Analysis of the microphone model

The voltage equation of the coil in Figure 8.19 is

$$v = \frac{d\lambda}{dt} + iR, \tag{8.24}$$

where $\lambda = \lambda(i, x)$ is the flux-linkage of the coil, R is its resistance, i is the current, and v is the terminal voltage. The flux-linkage is related to the current by the inductance L:

$$\lambda = L(x, i)i \tag{8.25}$$

where L is normally a function of both the airgap and the current (owing to saturation). However, if we assume that the iron permeability is effectively infinite, the dependence on current is eliminated and we can write

$$L = \frac{\mu_0 N^2 A}{2x} \tag{8.26}$$

where N is the number of turns on the coil, A is the airgap area, and μ_0 is the permeability of free space. Equations (8.25) and (8.26) can be expressed as

$$\lambda = \frac{ki}{x} \quad \text{where} \quad k = \frac{\mu_0 N^2 A}{2}. \tag{8.27}$$

The rate of change of flux-linkage $d\lambda/dt$ can be expanded by the chain rule, thus

$$\frac{d\lambda}{dt} = \frac{\partial\lambda}{\partial i}\frac{di}{dt} + \frac{\partial\lambda}{\partial x}\frac{dx}{dt} = v - iR. \tag{8.28}$$

From (8.27)

$$\frac{\partial\lambda}{\partial i} = \frac{k}{x} \quad \text{and} \quad \frac{\partial\lambda}{\partial x} = -\frac{ki}{x^2}. \tag{8.29}$$

Substituting (8.29) into (8.30) yields

$$\frac{k}{x}\frac{di}{dt} - \frac{ki}{x^2}\frac{dx}{dt} = v - iR. \tag{8.30}$$

Letting $u = dx/dt$ and rearranging,

$$\frac{di}{dt} = \frac{x}{k}v - \frac{xR}{k}i + \frac{i}{x}u. \tag{8.31}$$

The mechanical dynamic equation is

$$f_e = m\ddot{x} + b\dot{x} + k(x - x_0) \tag{8.32}$$

where f_e is the electromagnetic force, m is the relevant moving mass, b is the damping coefficient, k is the spring constant of the structure, and x_0 is a reference value of the displacement x. The force is given in [1] as

$$f_e = -\frac{ki^2}{2x^2}. \tag{8.33}$$

Substituting in (8.32) and rearranging,

$$m\ddot{x} = -b\dot{x} - K(x - x_0) - \frac{ki^2}{2x}. \tag{8.34}$$

The system can now be described by three nonlinear equations obtained from (8.31) and (8.34):

$$\frac{di}{dt} = \frac{x}{k}v - \frac{xR}{k}i + \frac{i}{x}u$$

$$\dot{x} = u$$

$$\dot{u} = -\frac{b}{m}u - \frac{K}{m}(x - x_0) - \frac{ki^2}{2x^2m}. \tag{8.35}$$

The nonlinear equations (8.35) can be linearized to provide insight into the electro-dynamic interactions. There are three states: current i, velocity u and displacement x. There is one input: voltage, v. The linear system is represented by writing $f_1 = di/dt$, $f_2 = \dot{x} = u$ and $f_3 = \dot{u}$, and computing the Jacobian matrix for both the states and the input: thus

$$\begin{bmatrix} f_1 \\ f_2 \\ f_3 \end{bmatrix} = \begin{bmatrix} \dfrac{\partial f_1}{\partial i} & \dfrac{\partial f_1}{\partial x} & \dfrac{\partial f_1}{\partial u} \\ \dfrac{\partial f_2}{\partial i} & \dfrac{\partial f_2}{\partial x} & \dfrac{\partial f_2}{\partial u} \\ \dfrac{\partial f_3}{\partial i} & \dfrac{\partial f_3}{\partial x} & \dfrac{\partial f_3}{\partial u} \end{bmatrix} \cdot \begin{bmatrix} i \\ x \\ u \end{bmatrix} + \begin{bmatrix} \dfrac{\partial f_1}{\partial v} \\ \dfrac{\partial f_2}{\partial v} \\ \dfrac{\partial f_3}{\partial v} \end{bmatrix} \cdot v. \tag{8.36}$$

The nonlinear equations are linearized about the operating point $(i, x, u, v) = (i_0, x_0, 0, v_0)$. So, for example, the coil has a small displacement, but since the velocity u is zero, the current is simply v/R.

The terms in the Jacobian matrix of (8.36) are

$$\left.\frac{\partial f_1}{\partial i}\right|_0 = \left.\left(\frac{-xR}{k} + \frac{u}{x}\right)\right|_0 = \frac{x_0}{k}\left(-R - \frac{ku_0}{x_0^2}\right) = -\frac{x_0 R}{k};$$

$$\left.\frac{\partial f_1}{\partial x}\right|_0 = \left.\left(\frac{v}{k} + \frac{Ri}{k} - \frac{i}{x^2}u\right)\right|_0 = \frac{1}{k}(v_0 - i_0 R) - \frac{i}{x^2}u_0 = 0;$$

$$\left.\frac{\partial f_1}{\partial u}\right|_0 = \left.\frac{i}{x}\right|_0 = \frac{i_0}{x_0}; \tag{8.37}$$

$$\left.\frac{\partial f_1}{\partial v}\right|_0 = \left.\frac{x}{k}\right|_0 = \frac{x_0}{k};$$

$$\frac{\partial f_2}{\partial i} = \frac{\partial f_2}{\partial u} = \frac{\partial f_2}{\partial v} = 0;$$

$$\frac{\partial f_2}{\partial x} = 1; \tag{8.38}$$

and

$$\left.\frac{\partial f_3}{\partial i}\right|_0 = \left.\frac{-ki}{x^2 m}\right|_0 = \frac{-ki_0}{x_0^2 m};$$

$$\frac{\partial f_3}{\partial x}\bigg|_0 = \left(-\frac{K}{m} + \frac{ki^2}{x^3m}\right)\bigg|_0 = \left(-\frac{K}{m} + \frac{ki_0^2}{x_0^3m}\right); \tag{8.39}$$

$$\frac{\partial f_3}{\partial u}\bigg|_0 = -\frac{b}{m};$$

$$\frac{\partial f_3}{\partial v}\bigg|_0 = 0.$$

Substituting the partial derivatives into (8.36) yields

$$\frac{di}{dt} = \left(\frac{-x_0 R}{k}\right)i + \left(\frac{i_0}{x_0}\right)u + \left(\frac{x_0}{k}\right)v;$$

$$\dot{x} = u; \tag{8.40}$$

$$\dot{v} = -\frac{ki_0}{x_0^2 m}i - \frac{b}{m}u - \left(\frac{K}{m} - \frac{ki_0^2}{x_0^3 m}\right)x.$$

Multiply the first of equations (8.40) by k/x_0 and the third equation by m:

$$\frac{k}{x_0}\frac{di}{dt} = -Ri + \left(\frac{ki_0}{x_0^2}\right)u + v;$$

$$\dot{x} = u; \tag{8.41}$$

$$m\dot{v} = -\frac{ki_0}{x_0^2}i - bu - \left(K - \frac{ki_0^2}{x_0^3}\right)x.$$

Define the following:

$$\text{Normal inductance } L_0 = \frac{k}{x_0};$$

$$\text{Equivalent stiffness } K_{eq} = K - \frac{ki_0^2}{x_0^3}; \tag{8.42}$$

$$\text{Microphone constant } k_1 = \frac{ki_0^2}{x_0^2}.$$

Then (8.41) becomes

$$L_0\frac{di}{dt} = -Ri + k_1 u + v;$$

$$\dot{x} = u; \tag{8.43}$$

$$m\dot{u} = -k_1 i - bu - K_{eq}x.$$

The microphone constant k_1 in (8.42) is the same as the constant K in (8.27).
Applying the Laplace transform to (8.43) we have

$$i(s) = \frac{1}{L_0 s + R}v(s) + \frac{sk_1}{L_0 s + R}x(s) \tag{8.44}$$

and

$$x(s) = \frac{-k_1}{ms^2 + bs + K_{eq}} i(s) = \frac{-k_1}{K_{eq}} \cdot \frac{K_{eq}/m}{s^2 + (b/m)s + K_{eq}/m} i(s). \tag{8.45}$$

Define the natural frequency ω_n and damping ζ as

$$\omega_n^2 = \frac{K_{eq}}{m}; \quad 2\zeta\omega_n = \frac{b}{m}; \tag{8.46}$$

then (8.45) becomes

$$x(s) = \frac{-k_1}{K_{eq}} \cdot \frac{\omega_n^2}{s^2 + 2\zeta\omega_n s + \omega_n^2} i(s). \tag{8.47}$$

Equations (8.44) and (8.47) can be written in block diagram form as shown in Figure 8.20.

As a practical matter, the constants in (8.42) can be determined from electromagnetic analysis of the motor, in conjunction with (8.29). If the flux-linkage is measured as a function of current, angle and airgap, each term of the equation can be computed. Measuring the flux as a function of airgap is admittedly difficult for a motor already constructed. However, this term can be computed using finite-element analysis. For example, the flux map is computed for the nominal airgap and then if the airgap is perturbed, the flux map is computed again. The difference between the flux-linkages obtained from the two cases, divided by the amount of perturbation, yields the microphone constant.

Equation (8.29) shows that the microphone constant is a function of current and angle. Therefore, the gain of the feedback path changes as the operating mode changes. In general, the 8/6 motor is very efficient at making noise in the aligned position, and almost unable to make noise in the unaligned position, implying that the microphone constant must be small at the unaligned position and large at the aligned position.

The 8/6 motor was modelled using the *PC-SRD* computer program to compute the flux-linkage plots (magnetization curves). The airgap was perturbed slightly and the flux-linkage plots recomputed. The partial derivative of the flux-linkage with respect to the airgap was computed as the numerical difference of the flux-linkage at each angle and current, divided by the airgap perturbation. This is plotted for a current of 50 A in Figure 8.21. The aligned angle is 0 and the unaligned angle is 180°. Note that the constant is highest at the aligned position and decreases to a small value at the unaligned position, as discussed above.

Another aspect of noise generation is that the force generated by the rotor must be able to excite the appropriate vibration modes. For an 8/6 motor in the aligned position, the current control is both capable of generating force and exciting the radial–axial mode – the mode that makes the motor an efficient speaker. These two effects contribute adversely to noise. Avoidance of anything that can excite the radial–axial mode and the forces that can excite it is required for quiet motor operation.

8.8.2 Analysis of Wu and Pollock's deadbeat control

Wu and Pollock noted that stepping the voltage at a time interval equal to half of the primary mode of the stator shell greatly reduced acoustic noise [2]. Their method

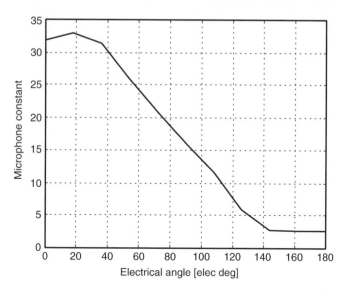

Fig. 8.21 Microphone constant for phase D at 50 A.

showed that voltage spikes directly affected the noise transmission; large current changes are not needed to induce noise. Additionally, it was noted that reducing the voltage steps at the aligned rotor position reduces the noise drastically compared to reducing the voltage steps at the unaligned rotor position. Both of these effects are consistent with the model of Figure 8.20.

Stepping the voltage is similar to the dead-beat control technique used in the control of flexible structures to minimize vibration. For example, suppose a long slender robotic arm is to be slewed with a force F. If half the force is initially applied at time zero, and the additional half force applied (total force of F) at a time equal to half the first mode of the arm, the first mode of the arm will not be excited. Similarly, in Wu and Pollock's method the voltage impulse is applied in two equal levels separated in time by one-half the natural frequency of the shell.

The fact that the effect of stepping the voltage is greater when the rotor is aligned than unaligned is also explained by the microphone model. The microphone constant is larger when there is current in the motor at the aligned position than when the current is zero at the unaligned position, allowing the voltage spikes to create acoustic noise.

The microphone model can be used to predict the experimental results of Wu and Pollock qualitatively. Suppose the motor shell has a natural frequency of 2500 Hz with a damping of 5%. Additionally, suppose that for a current of 80 A, the stator has a radial force of 5 kN and the stator is stiff enough so that such a force perturbs the shell by 1% of its nominal airgap. Let the nominal airgap be 0.25 mm. Figure 8.22 shows the velocity of the shell for both full step inputs in voltage and the case where the voltage is stepped by the Wu and Pollock method. By stepping the voltage, the shell vibration is greatly reduced. The results should be interpreted only qualitatively, but they do indicate that by avoiding voltage spikes that excite the radial axial mode, shell vibration and acoustic noise can be reduced.

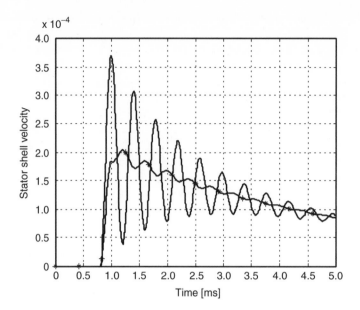

Fig. 8.22 Shell velocity for step voltage and double step voltage [2].

8.8.3   Current control loop topologies

Avoidance of voltage commands that excite the radial axial mode of the 8/6 motor is critical to reducing acoustic noise. Any control signal that excites the motor structure will induce noise. Proper shaping of the current control loop is one technique to reduce noise.

Figure 8.23 shows a block diagram of the current control topology used to reduce acoustic noise as described by Lu and McLaughlin (1998a,b). The current command is first passed through a notch filter to remove any unwanted frequency content. In the present application, the notch filter is realized as a digital version of the continuous filter,

$$G_{notch} = \frac{s^2 + 2(z/d)\omega_n + \omega_n^2}{s^2 + 2z\omega_n + \omega}$$

(8.48)

where the notch frequency is ω_n, the depth of the notch is determined by the value of d, and the width of the notch is determined by z. Increasing d increases the depth

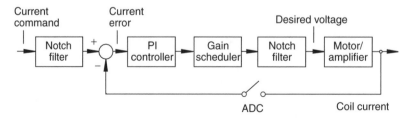

Fig. 8.23 Current control loop topology.

of the notch; decreasing z decreases the width of the notch. In the present application, $z = 0.707$, $d = 10$ and ω_n is set to the frequency of the radial axial mode of the motor stator. The notched current command is fed into a classic feedback control loop. The current error is computed and passed through a PI controller and then into a box termed the 'Gain Scheduler'. The gain scheduler compensates for the inductance variation in the motor as a function of current and angle, so that a constant-bandwidth current controller is achieved. If the inductance is low, as occurs at the unaligned rotor position or when the current is high, the gain is low; if the inductance is high, as occurs at the aligned position when current is low, then the gain is high. The output of the gain scheduler is fed into another notch filter to further remove frequency content that can excite the motor shell. The microphone model of the previous section is critical to the design of the gain scheduler and the notch filter. Using this model allows shaping of the current response transfer function and also the transfer functions that relate current command, voltage noise and current noise to acoustic noise. Also, linear stability analysis shows that if the notch filter is used inside the control loop, the gain scheduler is required to assure stability. Once the desired voltage is computed, it is formatted to drive the appropriate power switches of the amplifier.

A schematic of the power bridge is shown in Figure 8.24 (Miller, 1993). The motor phases are connected so that a single current sensor is shared between two phases; the selection of the lower switch determines which phase is driven. For a given phase, there are four possible operating modes as shown in Table 8.1.

The FETs are commanded such that at the beginning of the PWM cycle the bridge is in the upper recirculation mode with the upper FET closed and the lower FET open. If the voltage command to the phase is zero, then the amplifier will be in upper

Table 8.1 Operation Modes of Power Bridge

A_{upper}	A_{lower}	Mode
Closed	Closed	Charge
Closed	Open	Upper Recirculation
Open	Closed	Lower Recirculation
Open	Open	Recovery

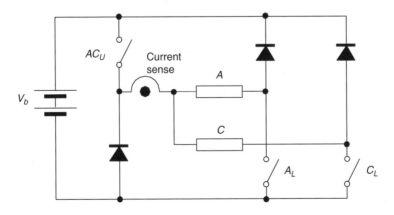

Fig. 8.24 Power amplifier topology.

recirculation for the first half of the PWM cycle and lower recirculation for the other half of the cycle. If the voltage command is greater than zero, the bridge begins in upper recirculation, then moves to charge mode, then to lower recirculation; the time in charge mode is equal to the desired voltage V divided by the battery voltage V_s times the PWM cycle time T_{chop} (see Sections 5.3.3 and 5.3.4): thus

$$V = \frac{t_{on}}{T_{chop}} \times V_s.$$ (8.49)

In the present example, the PWM cycle time is $T_{chop} = 42.5$ ms and the nominal battery voltage is $V_s = 13.5$ V. If a voltage of 3.5 V is desired across the coil (neglecting losses), then the bridge must be in charge mode for $(3.5/13.5 \times 42.5$ ms$) = 11.02$ ms; the bridge will be in upper recirculation for the first 15.74 ms, then in charge mode for 11.02 ms, then in lower recirculation for 15.74 ms. To achieve negative voltage across the coil, the same technique is used except that instead of being in charge mode in the middle of the PWM cycle, the bridge is commanded to recovery mode with both FETs open. The closed-loop nature of the loop automatically compensates for resistive losses.

The current control topology in conjunction with the power bridge allows precise control of the coil current without introducing voltage frequencies that can excite the acoustic producing modes of the stator structure. Analysis of the closed-loop control using the microphone model allows prediction of both current loop response and shell excitation.

8.9 Summary

Techniques were presented to achieve low torque ripple and low levels of acoustic noise in a switched reluctance motor with high torque density and exacting dynamic response requirements. Physical parameter knowledge of the motor is not required to obtain low torque ripple; however, it is required to reduce the acoustic noise. To reduce torque ripple, the motor/controller is viewed as a black box and the input command is modified to achieve the desired measured output. For low acoustic noise, the inductance map of the motor as a function of current and angle along with the frequency of the structure that dominates acoustic noise must be known. A model that demonstrates how the motor produces acoustic noise is presented and when this model is used to design a controller, both current control response and acoustic noise response can be predicted.

Drive development and test

Calum Cossar and Lynne Kelly

SPEED Laboratory, University of Glasgow and Motorola, East Kilbride

9.1 Introduction

As we have seen in previous chapters, the switched reluctance machine relies on control electronics not only to achieve basic rotation, but also to determine most of the important aspects of its performance, including the torque/speed range, the efficiency, the control loop bandwidth, the torque ripple, and the acoustic noise. At the present time there is no single 'off-the-shelf' switched reluctance controller, and the application-specific controllers used in various products are based on a range of technologies. In future it is anticipated that reductions in cost and increases in the functionality of the electronics will favour the switched reluctance machine and make it a more attractive option. A natural outcome of this trend could be the development of a highly integrated single-chip controller which will implement a range of control strategies at a cost which satisfies all but the lowest-cost, fractional-horsepower applications.

Given the wide range of applications for which the switched reluctance machine is either used or contemplated, each of the various performance features has to be assessed as to its cost/effectiveness. The most important specification issues are generally as follows:

- Number of quadrants of operation in the torque/speed diagram.
- Speed range: maximum speed, minimum speed.
- Braking requirements.
- Torque ripple.
- Acoustic noise.
- Variable speed control or servo-quality control.
- Position control.
- Efficiency optimization.
- Interface to complete system.

Before considering a range of implementations of switched reluctance controllers we will first of all review the various technologies which can be used, highlighting their advantages and disadvantages.

9.2 Implementation technologies

The potential options for the implementation of a switched reluctance controller can be categorized as follows:

- Analogue + discrete digital integrated circuits.
- Microcontroller.
- Digital Signal Processors (DSPs).
- Field Programmable Gate Arrays (FPGAs).

Analogue electronics (op-amps, comparators, etc.) together with a limited amount of discrete digital logic implemented in the TTL standard 7400 series was the traditional approach during the early 1970s as it was the only solution available to the designer. Used with power switching devices at relatively low switching frequency, these controllers tended to produce noisy operation which probably hindered the commercial development of the switched reluctance machine, even though its performance in other respects (especially efficiency) was competitive. Improved integration in manufacturing led to the development of a number of single-chip solutions arising from these early controllers, but the limitations remain, primarily in the optimized control of commutation throughout the torque/speed range.

The development of 8-bit microprocessors and microcontrollers during the late 1970s and early 1980s introduced a new era in real-time control implementation and these were applied at an early stage to exploit the potential of the switched reluctance machine (Chappell *et al.*, 1984). Considerable advances were made during this time in developing microcontroller-based drives, the primary aim of which was to develop optimized angle control throughout the complete torque/speed range. The microcontroller in particular is suited to this task, given its interrupt-driven topology and additional on-chip peripherals, especially the timer processing unit, analogue-to-digital converters, and serial communication channel. The nonlinear relationships between firing angles and speed and/or torque can be implemented in look-up tables, while the resultant commutation angles can be calculated in software and sent to the peripheral timer processing block within the microcontroller for real-time commutation control. The timer processing unit within the microcontroller is also useful in the calculation of actual motor speed. The analogue-to-digital converters can be used to input phase current sensors to allow software current regulation and interface to system control signals such as speed demand. A typical microcontroller is outlined in Figure 9.1. For more information on the advantages and implementation of microcontroller-based drive systems, see Bose (1986).

Despite the advances in control strategies, the microcontroller solution still tends to fall short of implementing 'high grade' inner-loop control strategies, due to the limited speed of software execution. Alternative technologies therefore had to be exploited, once they became available, to implement servo and sensorless control strategies.

Digital Signal Processors (DSPs) have been quick to make inroads in all areas of motor control due to their numerical processing capability. Combined with the later introduction of on-chip peripherals similar to those included in microcontrollers (TI TMS320C240, Analog Devices ADMC330), this has permitted 'single-chip' solutions

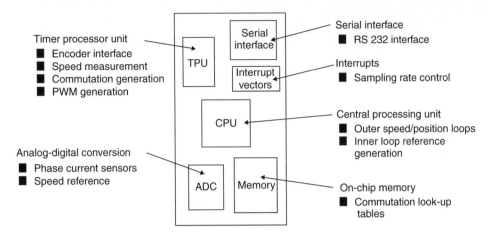

Fig. 9.1 Microcontroller outline.

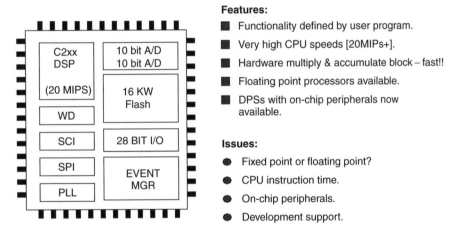

Fig. 9.2 DSP technology.

to a wide range of high-grade applications. Figure 9.2 outlines some of the main features and the issues associated with DSP development.

The alternative to using DSPs to implement fast, complex inner-loop control strategies is to use digital Application-Specific Integrated Circuits (ASICs), the most cost-effective solution for low/medium volume applications being Field Programmable Gate Arrays (FPGAs). A major advantage of FPGAs is that they are reprogrammable since they are configured by an associated memory device such as an EPROM. FPGAs allow the designer to create large application-specific digital circuits based on sequential and combinational logic elements which can be exploited to implement very fast inner-loop control topologies. However, FPGAs are less applicable to look-up table and outer-loop control requirements, so it is general to combine the FPGA with a micro-controller to implement a complete switched reluctance controller, as will be outlined later. Figure 9.3 outlines some of the features and issues associated with FPGAs.

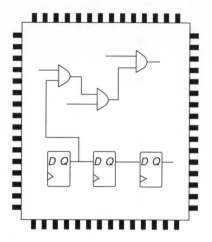

Features:

■ Digital hardware functionality defined by user.

■ Fast clock speeds [80 MHz] for sequential logic blocks.

■ Large circuits can be implemented on a single chip.

■ User reprogrammable versions available for prototyping [FPGAs].

Issues:

● Device manufacturer/family.

● User interface: VHDL, circuit diagrams.

Fig. 9.3 FPGA technology.

9.3   Functionality options

Given this brief outline of the various competing technologies we will now turn our attention to how each of these technologies can be applied to the control of switched reluctance motors. A typical control topology for a switched reluctance machine is shown in Figure 9.4. The options and issues associated with each control block are as follows.

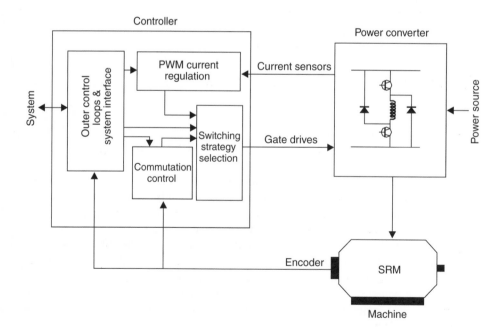

Fig. 9.4 Typical switched reluctance drive system.

Encoder

Issues:	
1	Resolution
2	Speed range
3	Environment
4	Cost

Options:	
1	Incremental encoder
2	Resolver
3	Hall effects
4	Sensorless

Commutation control

Issues:	
1	Speed range
2	Efficiency optimization
3	Resolution

Options:	
1	Fixed commutation angles
2	Commutation angles a function of motor speed
3	Angles a function of motor speed and torque
4	Zero voltage loop mode

Current regulation

Issues:	
1	Resolution
2	PWM frequency variation
3	Current ripple
4	Complexity
5	Number of current sensors
6	Speed range
7	Torque ripple

Options:	
1	Hysteresis control
2	Delta modulation
3	PI PWM current control
4	Voltage PWM
5	Single pulse mode
6	Instantaneous current regulation

Switching strategy selection

Issues:	
1	Current ripple
2	Braking required
3	Phaseleg losses
4	Instantaneous current control

(continued)

Options:	
1	Soft chopping
2	Hard chopping
3	Soft braking
4	Balanced chopping

Outer-loop control and system interface

Issues:	
1	Speed or position control
2	Variable speed or servo
3	System interface requirements
4	Instantaneous current control
Options:	
1	Two- or four-quadrant control
2	Analogue, digital and/or serial interface requirements
3	Outer-loop compensation algorithms: P, PI, PID

It is worthwhile noting at this stage that the final controller topology may need to include a range of options to meet the required specification, e.g. PI current regulation at low speed and single-pulse mode at high speed. Rather than examining each of these requirements in isolation, four controller topologies are now discussed to highlight the relationship between implementation complexity and functionality.

9.4 Application-specific controller examples

We will now consider four controller implementations which have different levels of sophistication:

1. Low-cost control chip based on analogue and discrete digital integrated circuits.
2. Microcontroller-based variable-speed drive.
3. DSP-based servo drive.
4. Microcontroller/DSP + digital FPGA high-speed position control actuator.

9.4.1 Low-cost control chip based on analogue + discrete digital integrated circuits

Given that switched reluctance machines compete with brushless d.c. motors in a number of applications some efforts have been made to develop a very low-cost single-chip solution to switched reluctance control along the same lines as brushless d.c. control chips (e.g. Motorola 33033). One such solution for a three-phase switched reluctance motor developed in the late 1980s is shown in Figure 9.5 (Miller, 1993).

The features of the controller are as follows:

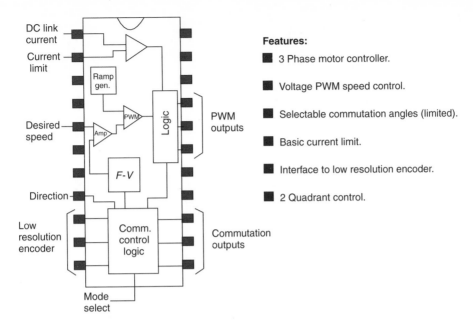

Features:

■ 3 Phase motor controller.

■ Voltage PWM speed control.

■ Selectable commutation angles (limited).

■ Basic current limit.

■ Interface to low resolution encoder.

■ 2 Quadrant control.

Fig. 9.5 Low-cost analogue + discrete digital controller.

- Operation from a low-resolution optical or Hall-effect encoder.
- Discrete firing-angle options.
- Voltage-PWM speed control.
- Overcurrent detection (hysteresis control).
- Soft chopping only.

The encoder pattern plus available commutation options (shown for phase 1 only) are shown in Figure 9.6. The desired commutation pattern is selected by connecting the associated digital control pins to a specific polarity.

Speed control is achieved by voltage-PWM control and is implemented in analogue electronics. Actual motor speed, determined by converting the frequency of one of the encoder channels to a proportional voltage, is compared with the analogue speed reference in an op-amp which outputs a resultant analogue error signal. This error signal along with a constant amplitude ramp signal is then input to a comparator the output of which becomes a pulse-width modulated signal whose frequency is set by the frequency of the ramp signal and whose duty cycle is proportional to the error signal. This PWM signal is then logically combined with each of the phase commutation signals to produce the relevant gate drives to one of the phaseleg switches (per phase). A diagram outlining this implementation is shown in Figure 9.7.

Overcurrent protection may be required at low speed and is implemented by an analogue hysteresis controller as shown in Figure 9.8. Hysteresis current control is the simplest to implement in analogue electronics, but results in variable switching frequency during the commutation period due to the variations in phase inductance and back-EMF.

The advantage of this controller is its low cost implementation, but performance is significantly compromised for the following reasons, so that this controller (or equivalent) could only be considered in very low-cost fractional horsepower applications:

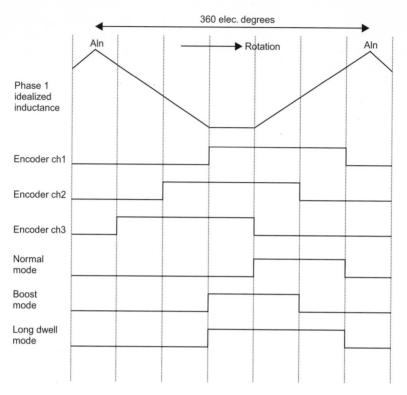

Fig. 9.6 Commutation outline.

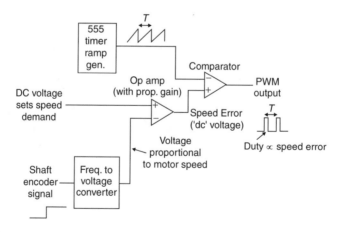

Fig. 9.7 Analogue speed control circuit.

- Drive efficiency not optimized throughout the torque/speed plane.
- 'Gear changing' of commutation angles required to utilize the complete torque/speed range of the motor. Additional electronics would be required to do this automatically.
- Two-quadrant control only (forward and reverse motoring).
- Poor low-speed performance (<60 rpm).

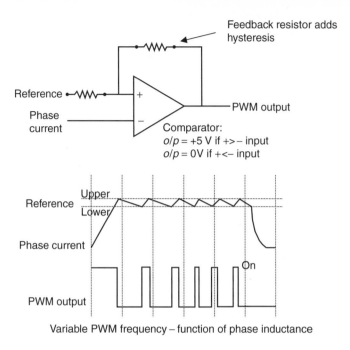

Fig. 9.8 Analogue hysteresis current control.

- PWM control mode only: therefore not optimized for high-speed operation (no single pulse speed control).

9.4.2   Microcontroller-based variable-speed drive

The most serious drawback of the previous implementation is without doubt the limitation in firing angle control which is essential for optimizing the performance of the switched reluctance drive in terms of efficiency, torque/speed envelope, torque ripple or acoustic noise. The introduction of the microprocessor and more recently the microcontroller was quickly identified by researchers in the late 1970s and 1980s as a cost-effective way of achieving optimized angle control over the complete torque/speed range.

Commutation angle control can be set as a function of either speed or speed and torque, depending on the application. In either case the nonlinear relationships are easily implemented in one- or two-dimensional look-up tables respectively, with additional interpolation to increase resolution performed in software. An example of such an implementation is outlined in Figure 9.9. The resultant commutation angles are converted to appropriate timer values before being sent by the CPU to the timer processing unit which performs the actual real-time commutation control for each motor phase. Typically the configuration of the timer/counter unit is flexible enough to allow interfacing to either high-resolution incremental encoders (counter function) or low-resolution Hall-effect sensors (timer function).

Figure 9.10 shows one example of how commutation control can be achieved with a high-resolution encoder. The commutation states are controlled as a function of encoder

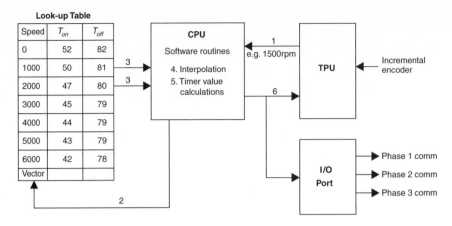

Fig. 9.9 Look-up table commutation angles vs. motor speed.

CHA transitions. This is best explained via an example.

Example: 1000-line encoder
                        Three-phase 6:4 machine
                        Commutation angles:
                        Turn-on angle $= 52°$ (mech.)
                        Turn-off angle $= 85°$ (mech.)

Assuming that the commutation look-up table software routine outputs the commutation angles in terms of encoder counts (144 and 236 in this example) then the next requirement is for the software to determine the required values to load into the counter in the correct sequence:

$$\text{Count value } X = T_{on} - 83 = 61$$

$$\text{Count value } Y = T_{off} - T_{on} - 83 = 9$$

$$\text{Count value } Z = T_{off} - T_{on} - 2.Y = 74$$

The sequence of events to derive the correct commutation pattern is then:

Time	Event	CPU action Comm state O/P PH3	PH2	PH1	CPU action Load counter value
t1	Index pulse interrupt	1	0	0	X (61)
t2	TPU counter $= 0$	1	1	0	Y (9)
t3	TPU counter $= 0$	0	1	0	Z (74)
t4	TPU counter $= 0$	0	1	1	Y (9)
t5	TPU counter $= 0$	0	0	1	Z (74)
t6	TPU counter $= 0$	1	0	1	Y (9)
t7	TPU counter $= 0$	1	0	0	Z (27)
t2	TPU counter $= 0$	1	1	0	Y (3)
etc.					

Whenever the required commutation angles change, as dictated by the look-up table, the count values X, Y and Z change accordingly. Note that for wide variations in

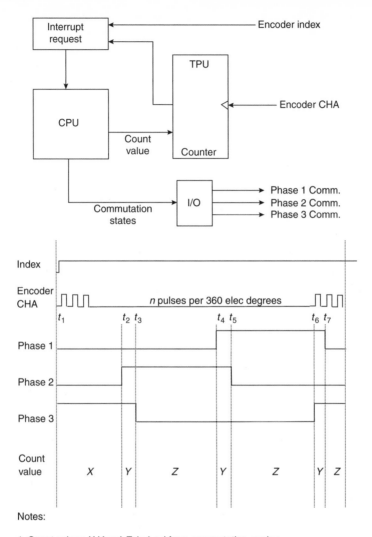

Fig. 9.10 Commutation control with high-resolution encoder.

commutation angles the sequence may not always be that described above; therefore an additional software routine is required to determine the required sequence for any particular value of commutation angles.

Interfacing with a low-resolution encoder places additional requirements on the microcontroller software and requires two TPU timer channels. Commutation points are now described in terms of TPU timer counts which are not only functions of the desired commutation angles (as with high-resolution encoders) but are also speed dependent. The software is once again required to determine timer values X, Y and Z and the subsequent sequence of events is similar to that described for the high-resolution encoder, but the calculation of X, Y and Z is more involved. Once again this is best

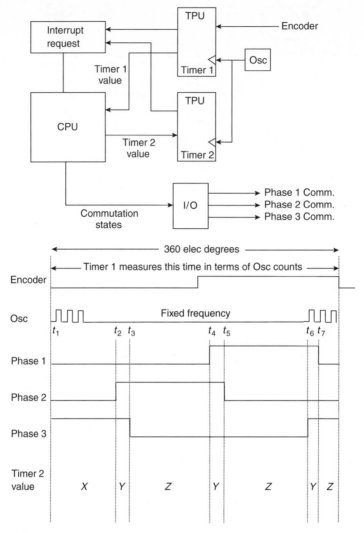

Notes:
1. Timer 2 values X, Y and Z derived from commutation angles and motor speed measurement (Timer 1 derived).
2. TPU Timer 2 is configured to count down.
3. Osc is fixed frequency oscillator within the microcontroller.

Fig. 9.11 Commutation control with low-resolution encoder.

described via an example (see Figure 9.11).

Example:	Motor speed:	1000 rpm
	Osc frequency:	1 MHz
	Motor:	Three-phase 6:4 ($360°$ (elec.) $= 90°$ (mech.))
	Turn-on angle:	$52°$ (mech.)
	Turn-off angle:	$85°$ (mech.)

Timer 1 is used to calculate motor speed in terms of osc periods. This is calculated as follows:

$$\text{Timer 1 value} = \frac{\text{Encoder period}}{\text{Osc period}} = \frac{2.39 \times 10^{-3}}{1 \times 10^{-6}} = 2390$$

This value of 2390 represents 90 electrical degrees, so the calculation of commutation points must use this as well as the desired commutation angles. The first stage in calculating X, Y and Z is to convert the commutation angles in terms of timer counts:

$$T_{on}(c) = \frac{52 \times 2390}{90} = 1380$$

$$T_{off}(c) = \frac{85 \times 2390}{90} = 2257$$

$$X = T_{on}(c) - H = 583 \qquad \left(\text{where } H = \frac{2390}{3} = 797\right)$$

$$Y = T_{off}(c) - T_{on}(c) - H = 80$$

$$Z = T_{off}(c) - T_{on}(c) - 2.Y = 717$$

Note that for a reasonable dynamic range in speed measurement a 16-bit timer is required with associated 16-bit arithmetic in code.

It is hoped that this gives some insight into microcontroller implementation of commutation control. The additional inner-loop requirement is that of phase current control which can either be implemented within the CPU code or using external hardware. An example of a low-cost/overhead control strategy favoured for digital implementation is Delta Modulation, the operation of which is shown for hardware implementation in Figure 9.12.

Mid-high-performance microcontrollers are also capable of implementing the outer speed control loop, where sample periods greater than 1 ms are generally used. A number of speed measurement techniques are possible using the microcontroller TPU, the most appropriate being a function of motor speed and encoder resolution. Considering interfacing to a high-resolution encoder, two strategies could be employed; the first

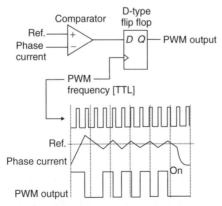

Fig. 9.12 Delta modulation current control.

counts the number of encoder CHA pulses within the speed loop sample period and is used at higher speeds; while the second measures the time period between successive encoder CHA edges and is used during low-speed operation. In both cases a 16-bit counter/timer is generally required to give reasonable dynamic range. These techniques are illustrated in Figures 9.13 and 9.14 respectively. In the classical nested control loop topology the resultant torque demand from this block can be used to control an inner torque loop parameter via a linearizing conversion block. Numerous methods of controlling the various inner-loop parameters have been reported, but a typical approach is to control current reference or voltage-PWM at low speed and turn-off angle at high speed.

A typical microcontroller-based variable-speed drive is shown in Figure 9.15.

The benefits of the implementation are:

- Programmable firing angles optimized for each motor through the use of look-up tables
- Four-quadrant control possible (motoring and braking)
- Software implementation of slower output speed and position control loops where necessary
- Highly integrated controller

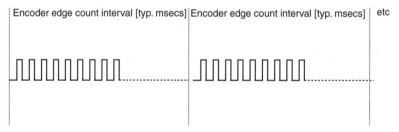

Notes:
1. Speed proportional to number of counts/sample period.
2. The number of edges detected defines the resolution of the speed measurement.

Fig. 9.13 High-speed measurement.

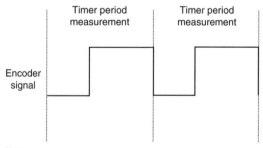

Notes:
1. Timer period measurement inversely proportional to motor speed.
2. Resolution of measurement is a function of timer number of bits and timer clock frequency.

Fig. 9.14 Low-speed measurement.

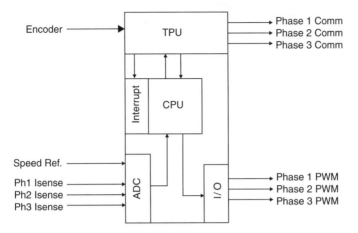

Fig. 9.15 Typical microcontroller-based variable-speed drive.

The limitations of the implementation are:

- CPU not fast enough (or there are better options) for applications which require 'instantaneous' high grade current control or sensorless operation.
- Update rate and high-speed accuracy of firing angles may be limited by the operation of the timer/counter unit. The user has to have very detailed knowledge of timer/counter unit operation before selecting microcontroller if high-speed operation is required (>10 000 rpm).
- 'Advanced' inner-loop control, e.g. PI current regulation, soft braking, may be too demanding for software implementation.

9.4.3 DSP-based servo drive

The exceptionally fast processor speeds plus the recent single-chip implementation of DSPs with peripherals such as ADCs, timer units, PWM generators and serial ports (e.g. TI TMS320C240, Analog Devices ADMC300) has resulted in significant application of this technology to all types of motor control. All the advantages of the previous implementation are applicable but because of the superior 'number crunching' a high grade (PI) instantaneous current regulator as required for servo operation can now also be implemented in software. Note that this feature also leads to DSP implementation of high-grade sensorless control strategies as outlined in Chapter 7. The reasons behind instantaneous current control have been discussed earlier, the aim being to 'profile' each phase current with respect to position and torque demand to minimize torque ripple. The implementation of this within the DSP CPU can be split into two parts both of which must be executed within one switching period (50 µsec for 20 kHz):

- Determine the present current reference for each phase.
- Calculation of new duty cycle for each phase according to a PI control algorithm.
- Output new duty values to on-chip PWM generators.

A block diagram of the required action is shown in Figure 9.16. There are numerous ways of implementing the PI current control strategy (Kjaer, 1997a,b), an example of

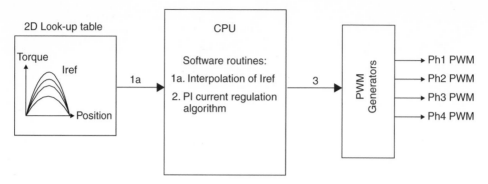

Instantaneous PI current regulation:

Step 1: Fetch instantaneous current reference

Step 2: PI current regulation algorithm

Step 3: Output new duty to PWM generator

Step 4: Repeat for all phases

(Performed every switching period e.g. 50 μs for 20 kHz PWM)

Fig. 9.16 Current profiling and PI current regulation.

which is as follows:

$$D_n = D_{n-1} + K_1 \cdot \varepsilon_n - K_2 \cdot \varepsilon_{n-1}$$

where D = duty, $\varepsilon = (I_{ref} - I_{act})$, K_1 = coefficient 1 and K_2 = coefficient 2.

The primary advantage of PI current regulation over the previously described hysteresis and delta modulation strategies is that the PWM chopping frequency is fixed. Additional benefits can also be gained by employing a double-sided modulation strategy in the PWM generation block (Blaabjerg *et al.*, 1999).

The advantages of the high processing speed and on-chip peripherals make the single-chip DSP + peripheral a very attractive option, but the following points should be considered before adopting this approach:

- The peripheral functionality is fixed in hardware, therefore potentially limiting the flexibility of the controller.
- Applying a combination of modes may require additional discrete digital hardware.
- High-speed machines have such low electrical time constants that very high-speed sampling of phase currents may be necessary for protection purposes – this may be better done using an FPGA implementation as outlined in option 4.
- Integrated versions are for fixed-point DSPs. The use of a floating-point processor for simplified program development (no scaling issues) would require additional peripheral devices for complete controller implementation (see option 4).
- CPU intensive – may become critical if the number and complexity of tasks become excessive.

Irfan Ahmed (1991) gives further reading in the application of DSPs to advanced motor control.

9.4.4   Microcontroller/DSP + digital FPGA high-speed position control actuator

This topology involves the partitioning of the control loops with a microcontroller or DSP implementing all the slower outer control loops and reference generation for all inner-loop control variables (plus other system level tasks), and a very fast and reprogrammable FPGA implementing all real-time inner-loop control and switching strategy selection. The result is a very powerful, flexible and 'future-proof' controller. An outline of the implementation of this control topology is shown in Figure 9.17. A critical feature of a high-speed actuator is the combination of low inductance and the requirement to operate at zero speed, which results in high current ripple and overcurrent conditions if advanced current regulation and protection strategies are not implemented. The advanced current regulation strategies require high phase current sampling rates (500 kHz+), and high-frequency/high-resolution PI current regulation (20 kHz/10 bit) which in turn demand very fast digital implementation in FPGAs. The inclusion of commutation control in the FPGA results in improved accuracy at high motor speeds due to reduced latency compared with microcontroller/DSP implementation. The implementation of soft braking also requires very fast current sampling and commutation/PWM control for low-inductance machines which again is best accomplished in an FPGA. The partitioning also results in the processor requirements being limited to software-only tasks (no on-chip peripheral requirements) the advantages of which are as follows:

- Easier upgrade path to newer processors.
- Conversion to floating point DSP with little change to FPGA implementation (assistance with determining motor speed may need to be incorporated into FPGA).

With the development of 'systems on a chip' with very high levels of integration, the realization of a high-performance processor plus flexible high-speed inner-loop control logic in a single chip could present a very attractive solution to all but the most cost-conscious applications.

9.4.5   Summary

As has been shown, the implementation of the switched reluctance controller is very much defined by the application requirements, resulting in a range of solutions. The following chart attempts to summarize the solutions against functionality requirements.

Controller #1:	Analogue + discrete digital
Controller #2:	Microcontroller
Controller #3:	DSP (integrated peripherals)
Controller #4:	Microcontroller/DSP + FPGA

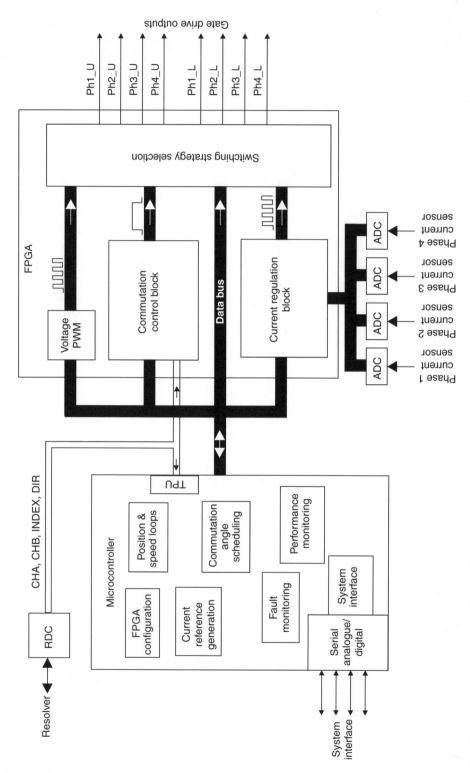

Fig. 9.17 Processor + FPGA controller topology.

Feature	#1	#2	#3	#4
Commutation control:				
Fixed/limited commutation angles	•			
Variable angle control with low-resolution encoder		•	•	
Variable angle control with high-resolution encoder		•	•	•
High-resolution/high-speed commutation control				•
Current control:				
Voltage-PWM	•	•	•	•
Hysteresis current control	•			
Delta modulation		•	•	•
PI current control			•	•
Current profiling			•	•
High-speed/high-resolution PI current control				•
High-speed current sampling				•
Outer loops:				
Variable speed control	•	•	•	•
Servo speed/position control			•	•
Sensorless control			•	•

9.5   Drive testing/optimization

Notwithstanding the considerable developments in switched reluctance machine simulation tools, there remains a need to prototype switched reluctance drives at an early stage in the development cycle, for the following reasons:

- Optimization may be a complex function of a number of criteria, some of which can only be established through experimentation, e.g. acoustic noise generation.
- As has been discussed earlier there is a wide range of controller implementations which results in potential cost savings if the minimum control strategy is identified early on.
- Validation of simulated performance predictions thus building up confidence in the simulation tool.
- Difficulty in accurately predicting magnetization curves in regions of heavy saturation, and differences due to manufacturing procedures.

The test requirements can be split into two distinct phases:

1. Measurement of machine magnetization curves.
2. Dynamometer tests of complete drive.

9.5.1   Measurement of magnetization curves

Magnetization curves define the relationship between flux-linkage and phase current for any rotor position, Figure 9.18. The measurement of magnetization curves for any particular machine is an important step in validating the performance of simulation tools and for determining optimized control strategies, such as current profiling to achieve smooth torque. There are two methods which can be used to measure these curves either directly or indirectly:

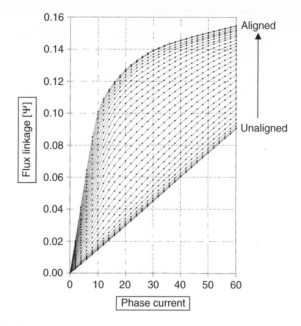

Fig. 9.18 Switched reluctance machine magnetization curves.

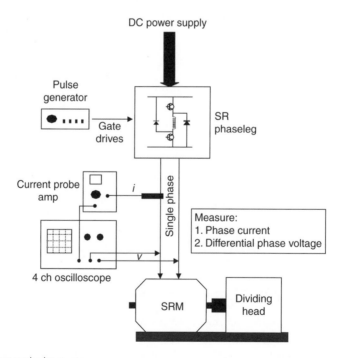

Fig. 9.19 Direct method test setup.

Test procedure:

1 Lock rotor at desired position using dividing head.

2 Energize phase winding using pulse generator.

3 Adjust frequency/duty cycle until the desired peak current is attained.

4 Sample phase current & voltage:

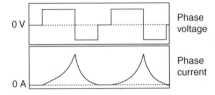

5 Process $v(t)$ and $i(t)$ to determine flux linkage:

$$\psi(t) = \int (v(t) - i(t).r).dt$$

Fig. 9.20 Direct method test procedure.

The direct method

The test setup is shown in Figures 9.19 and 9.20. The processing required to be carried out either on the data acquisition system or uploaded to a PC is given in Figure 9.21.

The indirect method

This is based on the measurement of static torque curves as a function of phase current and rotor position (Figure 9.22) and then converting this data to the required magnetization curve format as outlined in Figure 9.23.

Cossar and Miller (1992) give a complete outline of these test methods and a comparison of results.

9.5.2 Dynamometer tests of the complete drive system

Given the wide range of control mode options available and the difficulty in determining the optimized control strategy through simulation, there is a need for a *flexible controller* which has the following features.

No.	Specification	Example
1	Direct user access to all inner loop control variables	commutation angles current reference switching strategy switching frequency
2	Wide range of control mode options	speed control position control PI current regulation current profiling delta modulation soft braking angle control tables

(continued)

No.	Specification	Example
3	User-friendly interface	PC control panel (Windows)
4	Close integration with simulation packages	*PC-SRD*
5	Interface to a wide range of encoders	incremental encoders (all resolutions)
		Hall-effect sensors and resolvers
6	Built-in data acquisition system	performance variables can be displayed on PC
7	Large number of analogue and digital I/O	
8	Control of sampling rates and the resolution of a number of variables	commutation angles current reference
9	Faulted and unbalanced modes	shorted phaseleg switch simulation unbalanced phase currents
10	Full user access to all control and performance variables	

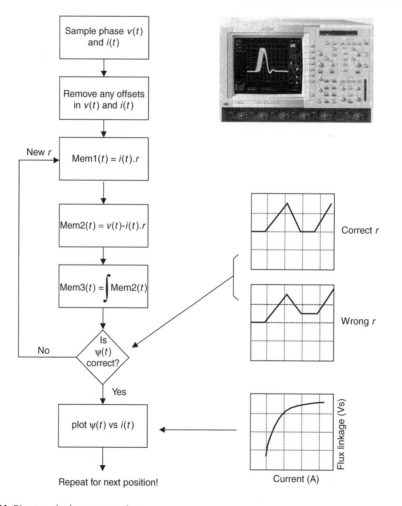

Fig. 9.21 Direst method post processing.

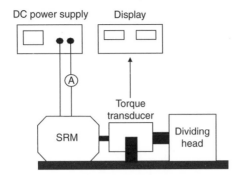

Fig. 9.22 Indirect method test setup.

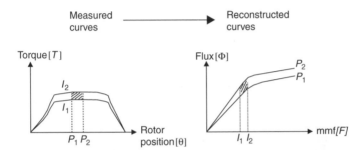

Measured curves ⟶ Reconstructed curves

Areas represent the same work done which can be represented mathematically as follows:

$$\int_{\theta_n}^{\theta_{n+1}} (T_{F_{n+1}} - T_{F_n})\,\delta\theta \quad = \quad \int_{F_n}^{F_{n+1}} (\Phi_{\theta_{n+1}} - \Phi_{\theta_n})\,\delta F$$

■ The reconstruction process requires as an initial condition the knowledge of one measured magnetization curve – this is generally chosen to be the unaligned curve as it can be determined by a simple a.c. impedance test.

Fig. 9.23 Indirect method test procedure.

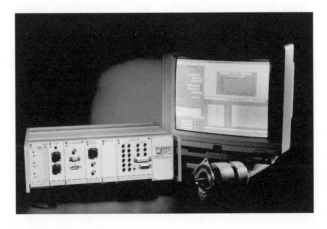

Fig. 9.24 University of Glasgow Flexible Controller FCIII.

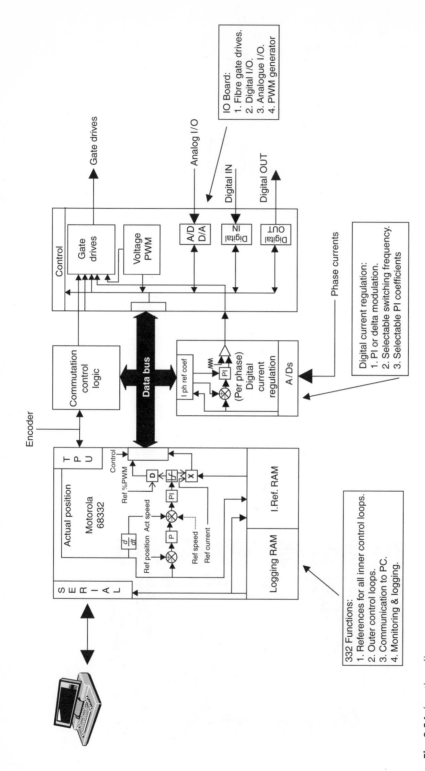

Fig. 9.24 (*continued*).

Using the information gathered through dynamometer tests, the lowest-cost implementation of the required strategy can then be developed with a high degree of
confidence. Figure 9.24 shows an example of such a flexible controller developed
at the University of Glasgow (Kelly *et al.*, 1999). The implementation is based on a
microcontroller + FPGA topology as outlined earlier.

An example of the type of tests carried out on this system would be to determine
the optimized efficiency commutation angles in single-pulse mode at a number of
torque/speed values; see Figure 9.25.

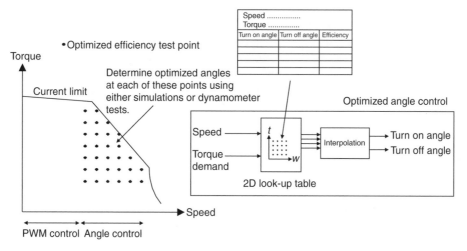

Fig. 9.25 Optimized efficiency tests.

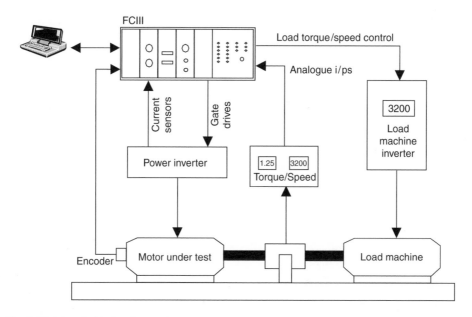

Fig. 9.26 Integrated test environment.

Taking the concept a stage further would be the development of a highly integrated test environment as shown in Figure 9.26, whereby the switched reluctance controller also controls either the torque or speed of the load machine (as a function of speed or position; for example, to mimic a certain load), and inputs external monitoring signals such as machine torque and temperature.

9.6 Summary

This chapter is concerned with the hardware of the controller and with the testing process. Several implementation options for the controller hardware are described and evaluated, including discrete logic, microcontrollers, field-programmable gate arrays, and digital signal processors. The chapter compares their functional ability in the realization of different levels of control. In the testing process, the measurement of magnetization curves is described, followed by the general approach to dynamometer testing.

10

The switched reluctance generator

Tadashi Sawata

10.1 Principle

The switched reluctance machine can operate as a generator as well as a motor by simply changing the firing angles. Figure 10.1 shows generating and motoring currents controlled by current chopping, together with the phase inductance.

In generating operation the firing angles are chosen so that the current flows when $dL/d\theta < 0$; while in motoring operation they are chosen so that the current flows when $dL/d\theta > 0$. The circuit equation for a phase of the switched reluctance machine is that:

$$v = Ri + \frac{d\phi}{dt}$$

$$= Ri + L\frac{di}{dt} + i\frac{d\theta}{dt}\frac{dL}{d\theta}$$

$$= Ri + L\frac{di}{dt} + e \tag{10.1}$$

where v is the applied voltage, i is the phase current, R is the phase resistance, L is the phase inductance and θ is the rotor position. The back-EMF e is defined as

$$e = \omega i\frac{dL}{d\theta} \tag{10.2}$$

where ω is the rotor speed $\omega = d\theta/dt$. As unipolar currents are normally employed the sign of i is always positive. Therefore, the sign of e is determined by $dL/d\theta$. When $dL/d\theta > 0$ the back-EMF is positive and it tends to force the current to decrease, being against the applied voltage and changing the electrical power supplied into mechanical output (motoring). When $dL/d\theta < 0$ the back-EMF is negative and it tends to increase the current and convert the mechanical power into electrical power (generating). However, the amplitude of the back-EMF varies with the rotor speed ω and the behaviour of the current is determined by the relationship between e and v, so that the actual operation is more complicated as described later.

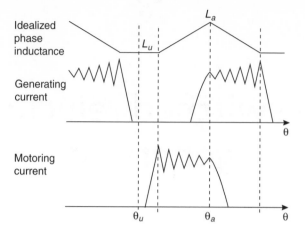

Fig. 10.1 Phase inductance and generating current.

Figure 10.2 shows the circuit diagram of a generator with one phaseleg and Figure 10.3 shows the phase currents, flux-linkage and idealized inductance. Both the converter and load are connected to the same d.c.-bus. The bus can be separated into two: one for the excitation and the other for the load for higher fault-tolerance (Radun, 1994).

A simplified circuit diagram showing the energy flow is shown in Figure 10.4. The integral of the currents in Figure 10.4 can be defined by referring to Figures 10.2 and Figure 10.3 as:

$$I_{in} = \int_{\theta_{on}}^{\theta_{off}} i_{ph} \, d\theta$$

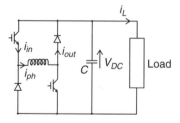

Fig. 10.2 Generator circuit for one phase. (Courtesy of Dr P.C. Kjaer).

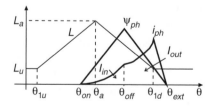

Fig. 10.3 Generator phase currents, flux-linkage and idealized inductance. (Courtesy of Dr P.C. Kjaer).

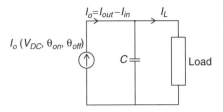

Fig. 10.4 Simplified circuit diagram showing energy flow. (Courtesy of Dr P.C. Kjaer).

$$I_{out} = \int_{\theta_{off}}^{\theta_{ext}} i_{ph} \, d\theta$$

$$I_o = I_{out} - I_{in}$$

where I_o is the net generated current. The excitation penalty ε is defined as follows:

$$\varepsilon = \frac{I_{in}}{I_{out}} = \frac{I_{in}}{I_o + I_{in}}. \tag{10.3}$$

An example of idealized current waveforms with single-pulse control is shown in Figure 10.5. The angles are defined in Table 10.1. The peak current occurs either at θ_{off} or θ_{1d}. Figure 10.5(a) shows the case where the current increases after turning off the switches at θ_{off}, when the back-EMF in the coil is larger than the d.c.-bus voltage V_{DC}. In (b), the back-EMF and V_{DC} balance and the current stays constant until the pole overlap ends at θ_{1d}. In (c) the back-EMF is smaller than V_{DC} and the current decreases after θ_{off}.

From θ_{on} to θ_{off} excitation power is supplied from the d.c. power source through the power electronic converter to the machine, and it is stored in the airgap as magnetic energy. After the power electronic switches are turned off at θ_{off} regenerated current keeps flowing through the freewheeling diodes returning the generated power into

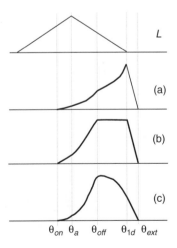

Fig. 10.5 Idealized current waveforms.

Table 10.1 Definition of the angles

θ_{on}	Turn-on angle
θ_a	Aligned position
θ_{off}	Turn-off angle
θ_{1d}	Angle at which pole overlaps ends
θ_{ext}	Angle at which the flux reaches zero

the d.c. power supply until the current vanishes at θ_{ext}. If the generated power P_{gen} is larger than the excitation power supplied from the d.c. supply P_{exc}, the system has *generated* the net power by converting the mechanical power into electrical power.

As generated power is returned to the d.c. power supply and/or filter capacitor the switched reluctance machine is a *d.c.* power generator. If the power is supplied to loads which require a.c. power, the d.c. power has to be converted into a.c. power by means of an inverter. In many applications all or part of the electric power generated has to be stored in batteries or other energy storage devices for a certain period of time. The switched reluctance generator may be able to charge the batteries directly through its power electronic converter for those applications.

The waveform of (a) has the smallest ε and (c) has the largest. If the net generated current is the same for all three cases, the current waveform of (a) is preferred since its smaller excitation penalty reduces losses.

Figure 10.6 shows the loci of the currents on the $i - \psi$ plane when the peak values are the same. The energy converted is proportional to the area enclosed by the loci. When the peak value of the current is limited, (b) is expected to generate the largest energy. Figure 10.7 shows typical simulated current waveforms for a speed of 10 000 rpm (Miller and McGilp, 1990). A prototype 0.8 kW single-phase 8/8-pole switched reluctance machine was used for the simulation, with phase resistance = 0.016 Ω and aligned/unaligned inductance = 1.94/0.21 mH. Figure 10.8 shows the loci of the currents on the $i - \psi$ plane together with measured magnetization curves. Three cases are shown with V_{DC} = 100, 270 and 350 V. The turn-off angles were adjusted to limit the peak value of the current to around 65 A while the turn-on angles were kept almost constant. The conditions and data calculated are summarized in Table 10.2,

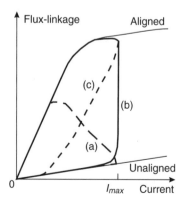

Fig. 10.6 Energy-conversion loop.

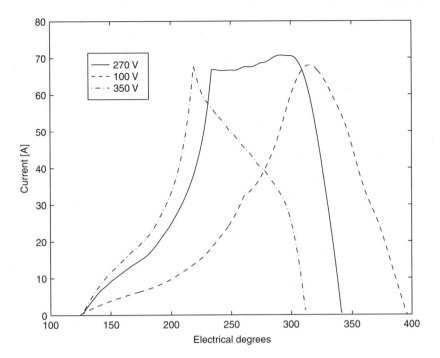

Fig. 10.7 Example of current waveforms.

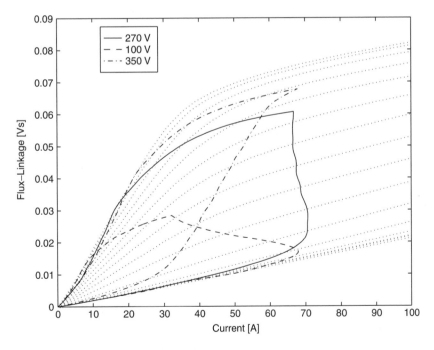

Fig. 10.8 Example of energy-conversion loops.

Table 10.2 Comparison of waveforms

V_{DC}	100 V	270 V	350 V
ε	0.256	0.350	0.525
I_o	12.0 A	11.5 A	5.2 A
P_o	1.20 kW	3.11 kW	1.81 kW
θ_{on}	124.0°	126.0°	126.0°
θ_{off}	260.0°	233.5°	218.5°

where I_o is the net generated current, and P_o is the power generated. The switching angles are shown in electrical degrees.

The current at $V_{DC} = 100$ V corresponds to (a), 270 V to (b) and 350 V to (c) in Figure 10.5. The excitation penalty with $V_{DC} = 100$ V is 0.256 and approximately half that with 350 V. The power generated with $V_{DC} = 270$ V is 3.11 kW, which is 2.6 times that with 100 V and 1.7 times that with 350 V. This is apparent from Figure 10.8. The case of $V_{DC} = 270$ V utilizes most of the possible energy conversion with the given current (65 A), while the others do only part of that.

If R is negligible equation (10.1) is rewritten as:

$$V_{DC} = L\frac{di}{dt} + e \qquad (10.4)$$

where $v = V_{DC}$. The speed at which $e = V_{DC}$ with the rated current is called *the base speed* (Miller, 1993). If V_{DC} is controlled as $V_{DC} = e$ the current keeps a constant value from the period from θ_{off} to θ_{1d} provided $\theta_{off} > \theta_a$. A control scheme which regulates V_{DC} with speed in order to maximize the energy conversion effectiveness by keeping the condition of $V_{DC} = e$ was proposed in Sawata *et al.* (1998).

The current waveform of switched reluctance generator may be controlled as follows:

- The current waveform should resemble (a) in Figure 10.5 when the generating current is small and the peak current, which is limited by the system, is less than its maximum value in order to minimize the losses. V_{DC} may be maintained at a nominal value.
- The current waveform should resemble (b) after the peak current reaches its maximum value in order to utilize the maximum energy available for a given maximum current. V_{DC} should be regulated to keep $V_{DC} = e$.

10.2　Review of switched reluctance generator control

The switched reluctance machine has been studied for generator applications especially where robustness, high speed and fault tolerance are important (Ferreira *et al.*, 1995; MacMinn and Jones, 1989; Radun, 1994; Richter and Ferreira, 1995; Richter, 1988). It has also been studied for wind energy generation (Fulton and Randall, 1986; Torrey and Hassanin, 1995), in which the speed range is wide and the operating speed is in the lower part of it for the majority of the time (Warne and Calnan, 1977; Buehring and Freris, 1981).

The objective of the generator control is normally to keep the d.c.-bus voltage at a desired value with the maximum efficiency. Figure 10.9 shows a typical switched reluctance generator control system. The d.c.-bus voltage V_{DC} is fed back and compared

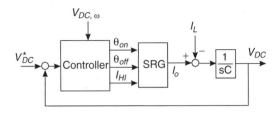

Fig. 10.9 Generator control system.

to its reference value V_{DC}^*. The controller regulates control variables such as the turn-on/off angles θ_{on}/θ_{off} and current set point I_{HI}. The controller may use V_{DC} and the rotor speed ω as input parameters. The SRG, which represents the converter and machine in Figure 10.9, generates current I_o. The difference between I_o and the load current I_L charges the filter capacitor C, the charge of which determines V_{DC}.

The necessity for a controller for the switched reluctance generator was shown by Radun. He showed in Radun (1994) analytically that the switched reluctance generator with fixed firing angles was unstable and the d.c.-bus voltage may increase or decrease exponentially depending on the load.

Relatively little has been reported on the design of the switched reluctance generator controller. MacMinn (1981) proposed a control scheme in which the turn-off occurs when the phase current reaches a 'reference turn-off value' and the turn-on angle is controlled linearly with the difference between the reference value and actual value of the d.c.-bus voltage (Figure 10.10(a)). He showed the linearity between the turn-on angle and generated current by simulation and experiment. When the requirement of generating power is low another mode of control is introduced. This is because the scheme keeps the current high even when the generating current required is small and this leads to reduced efficiency. In the low-power mode, the turn-on angle is fixed and dwell angle is controlled for the required output. A system efficiency of over 80% was reported at a speed of 47 000 rpm and generating power of over 20 kW. Experimental results of the regulator's response to step changes in load were also reported. A good linear relationship seems to be achieved with the scheme. Determining the turn-off current levels and turn-on angles for required generating power is the key element of the scheme, but this does not appear to be an easy task.

Ferreira *et al.* (1995) controlled the output power by regulating the phase current level with fixed turn-on and turn-off angles. The control is similar to the current chopping control in motoring mode; both power switches are on when the current is

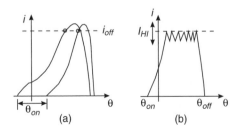

Fig. 10 Generator controls.

below the threshold and the switches are off when the current is above the threshold (Figure 10.10b). The threshold is the output of the PI controller of the d.c.-bus voltage. They reported that if only the angle control was used, the output current is very sensitive to small changes in the turn-on and turn-off angles and poor voltage control would result in practice. A good linear relationship between the current command and the output power was shown. A system efficiency of over 85% was reported at 41 000 rpm for the load of 20 to 30 kW. This scheme is relatively simple to implement but the relationship between the threshold of the phase current and the output power must be obtained by experiment or simulation prior to implementation.

Cameron and Lang (1993) studied the control of a switched reluctance generator system in electric power systems by simulation. The power system includes the generator and converter, a filtering capacitor, a distribution network and a load. The network comprises a series resistor and inductor, while the load comprises a parallel resistor, capacitor and power sink. The average output current as a function of the turn-on and turn-off angles was computed and mapped for the d.c.-bus voltage by simulation. For the d.c.-bus controller the turn-on angle is fixed and the turn-off angle is chosen to produce a desired current. The key feature here is that the d.c.-bus voltage controller is designed to include the dynamics of the power system network.

Radun (1993) proposed a scheme to linearize the relationship between the switching angles and the average generated current. When the d.c.-bus voltage V_{DC} increases the average generated current increases with constant switching angles. This nonlinearity is compensated for by introducing the quantity of V_{DC}/V_0, where V_0 represents the nominal d.c.-bus voltage. The turn-on and turn-off angles, that are determined from the function $F(\theta_{on}, \theta_{off})$, are multiplied by V_{DC}/V_0:

$$I_{ave} = \frac{V_{DC}}{V_0} F(\theta_{on}, \theta_{off}).$$ (10.5)

With this scheme, it was claimed that a linearized relationship between the switching angles and the average generated current was obtained over a wide range of the d.c.-bus voltage.

The use of soft-chopping was proposed by Stephens and Radun (1992) to control the current of the switched reluctance generator at relatively low and medium speeds where the current decreases with the application of negative d.c. voltage. Both switches are turned on at the turn-on angle and remain on until the current reaches the upper hysteresis level. When the current reaches the upper hysteresis level both switches are turned off and the current decreases. One switch is turned on when the current reaches the lower hysteresis level and the current freewheels through that device and the corresponding freewheeling diode. The process repeats until the turn-off angle is reached. This scheme is expected to reduce chopping frequency and hence reduce switching losses and result in lower current ripple.

Kjaer et al. (1994) derived an 'inverse model' of the switched reluctance generator which gives an analytical expression for the generated power. It was shown that the amount of energy generated is determined only by the turn-off angle and the turn-off current level at a given speed and d.c.-link voltage, and that chopping is not necessary. The limitation here is that the analysis assumed linear magnetic characteristics in the machine.

Kjaer *et al.* (1995) also discussed a technique to improve the efficiency of the switched reluctance generator by on-line reduction of the reactive power flow. In steady state, the turn-off angle is used to seek an optimum operating point while the d.c.-link voltage is regulated at a reference value by changing the turn-on angle.

All work described above assumes that V_{DC} is controlled to a constant value. On the other hand, a constant V_{DC} will limit the speed range of the generating operation since the back-EMF is a function of speed. Some method of control which enables wide speed range operation may be required especially for applications where the speed of the prime mover varies over a wide range.

10.3 Control method for a higher energy conversion effectiveness

One way to achieve maximum energy conversion with the switched reluctance generator system is to regulate the d.c.-bus voltage with the rotor speed. The drawback of the scheme is the necessity of an extra voltage regulator or d.c./d.c. converter between the system and a constant voltage load. However, a simple step-down converter may be sufficient as shown later.

Figure 10.11 shows the generator converter with an additional step-down converter proposed in Sawata *et al.*, (1998). The excitation energy is supplied from the capacitor C during the excitation period. During the generation period the generated energy is stored in C. When the generator starts operation with the capacitor discharged the excitation energy may be supplied from an external voltage source V_{ext} until the d.c.-bus voltage is established. The d.c.-bus voltage V_{DC} across C is regulated with speed for the ideal 'flat-top' shape by a d.c.-bus voltage controller. The load voltage V_L is regulated at a constant value by chopping Q_3 by a load voltage controller.

Figure 10.12 shows the block diagram of the d.c.-bus voltage controller. The DC-bus voltage command V_{DC}^* is calculated from the speed and then compared with its actual value. The PI controller outputs the reference value of the net generating current I_o^* from the d.c.-bus voltage error. The turn-off angle θ_{off} is controlled linearly with I_o. The turn-on angle θ_{on} is determined by the inverse model using θ_{off} and I_o.

Figures 10.13 and 10.14 show the phase currents at speeds of 3000 and 10 000 rpm. The d.c.-bus voltage is regulated at 89 and 297 V respectively. Both currents have more or less the 'flat-top' shape. The following parameters were used here: a single-phase 8/8-pole machine (0.8 kW), a chopping frequency of 5 kHz for Q_3, $C = 4500\,\mu F$, $L = 10\,mH$ and a resistive load of 20 Ω.

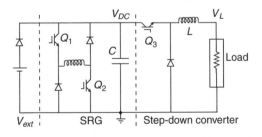

Fig. 10.11 Generator converter with an additional step-down converter.

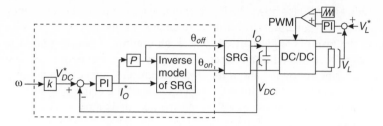

Fig. 10.12 Proposed generator controller.

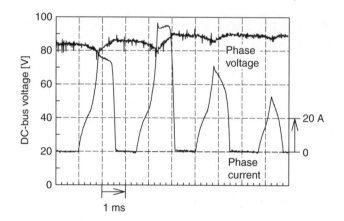

Fig. 10.13 Ripple in d.c. bus voltage.

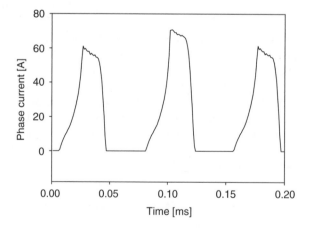

Fig. 10.14 Current waveform at 10 000 rpm.

Figure 10.13 also shows the DC-bus voltage for a speed of 3000 rpm. It can be seen that the d.c.-bus voltage is regulated at the command value of 89 V by varying the generating current. The relatively large ripple in the d.c.-bus voltage shown can be reduced with a larger filter capacitor at the expense of lower dynamic response and higher cost. The use of a poly-phase machine is also effective in reducing the ripple.

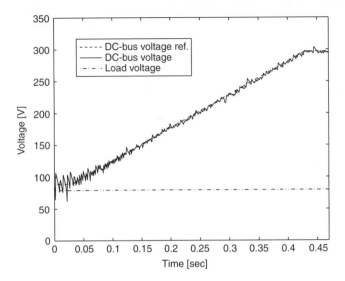

Fig. 10.15 Change in the d.c.-bus voltage with speed.

Figure 10.15 shows how the d.c.-bus voltage changes with the rotor speed. The rotor speed was changed from 3000 rpm to 10 000 rpm. The d.c.-bus voltage changes proportionally to the speed from 89 to 297 V while the load voltage is kept constant at 80 V by the step-down d.c./d.c. converter connected between the generator converter and the load. This shows that the d.c.-bus voltage can be controlled dynamically and independently of the load voltage with the system. A simple step-down d.c./d.c. converter is sufficient for the system provided that the minimum d.c.-bus voltage is higher than the load voltage over the range of operating speed.

10.4 PWM control

Voltage-PWM control can be used to regulate the current to a desired value by varying the average applied voltage in the switched reluctance generator. However, many applications in which the generator is likely to be used operate at very high speed which may make the application of PWM difficult. Another disadvantage of using PWM is that the excitation penalty is higher and the operational speed range is narrow. This is analysed in the following. In the analysis the semiconductor switches and diodes are assumed to be ideal and the voltage drops in those devices are neglected. The inductance profile is assumed to be linear. It is assumed that the turn-on is just before the aligned position and the current reaches its set point in the aligned position. The turn-off angle is assumed to be in the position where the inductance reaches its minimum value.

In one PWM period the excitation current corresponds to the area A_1 and the generated current corresponds to A_2 in Figure 10.16. Assuming that the duty cycle d is constant over a commutation period and the current ripple due to PWM is negligible compared to its average peak value i_p,

$$A_1 = D_1 t_{pwm} i_p$$

$$A_2 = D_2 t_{pwm} i_p \tag{10.6}$$

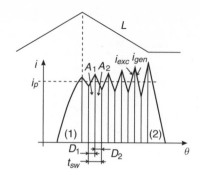

Fig. 10.16 PWM control.

where $D_1 = d$, $D_2 = 1 - d$ and t_{pwm} is the inverse of PWM frequency $t_{pwm} = 1/f_{pwm}$. The net generated current in one PWM period $I_{o,pwm}$ is:

$$I_{o,pwm} = A_2 - A_1$$
$$= (D_2 - D_1)t_{pwm}i_p \qquad (10.7)$$

In the analysis, the initial excitation current (area indicated as (1) in Figure 10.16) and the generated current after the turn-off angle (area indicated as (2)) are neglected. The ratio between excitation energy and generated energy ξ for the flat-top current with the PWM is defined as:

$$\xi = \frac{D_1}{D_2 - D_1} \qquad (10.8)$$

$$= \frac{d}{1 - 2d}. \qquad (10.9)$$

During the excitation period, the circuit equation for a phase is expressed as follows:

$$V_{DC} = L\frac{di_{exc}}{dt} - e \qquad (10.10)$$

where the resistance of the coil is neglected and the back-EMF e is

$$e = i \left| \frac{dL}{d\theta} \right| \omega \qquad (10.11)$$

where ω is the rotor speed. For generating operation $dL/d\theta$ is always negative and for clarity the negative sign is placed outside e. It should be noted that for PWM control the condition of $|V_{DC}| > |e|$ must be always true otherwise the generated current is uncontrollable and increases during the freewheeling period. Since the current increases during the excitation period,

$$L\frac{di_{exc}}{dt} = V_{DC} + e > 0. \qquad (10.12)$$

During the generating period, the equation is:

$$-V_{DC} = L\frac{di_{gen}}{dt} - e \qquad (10.13)$$

and

$$L\frac{di_{gen}}{dt} = -V_{DC} + e < 0. \tag{10.14}$$

Assuming that the change in L in a PWM period is small and neglected, in order to keep the average current constant over a PWM period,

$$\left|\frac{di_{exc}}{dt} \times D_1 t_{pwm}\right| = \left|\frac{di_{gen}}{dt} \times D_2 t_{pwm}\right|$$

$$|V_{DC} + e| \ D_1 = |-V_{DC} + e| \ D_2$$

$$(V_{DC} + e) \ D_1 = (V_{DC} - e) \ D_2.$$

By rearranging the equation above we obtain:

$$d = \frac{1}{2}\left(1 - \frac{e}{V_{DC}}\right). \tag{10.15}$$

From equation (10.9) if $d \geq 0.333$ then $\xi \geq 1$. In order to generate net energy which is greater than the excitation energy the condition of $d < 0.333$ must be satisfied. Substituting this condition into equation (10.15) gives the condition of $e > 0.333\,V_{DC}$. Therefore, the condition to generate net power more than the power used for excitation with PWM control is:

$$0.333V_{DC} < e < V_{DC} \tag{10.16}$$

and

$$0.333\omega_B < \omega < \omega_B \tag{10.17}$$

where ω_B is the base speed.

The result indicates that to generate net power more than the excitation power with PWM control the rotor speed must be higher than a third of the base speed and less than the base speed (speed range of 1:3). For applications which require wider operating speed range, PWM control does not appear to be suitable.

10.5   Design of a controller for the switched reluctance generator

The I_o and ε for a set of turn-on and -off angles for a single-phase 8/8 switched reluctance machine are shown in Figure 10.17 and 10.18 respectively. These were calculated by using the inverse machine model proposed by Kjaer et al. (1994a).[1] The main task in designing a generator controller is to obtain the relationships between the switching angles and/or current set-point (controller input) and generating power (controller output) for a constant d.c.-bus voltage (controlling the plant output for a reference). This is necessary regardless of the type of controllers employed. The mapping between

---

[1] Figure 10.17 indicates the trend that I_o increases by advancing the turn-on angle and delaying the turn-off angle. It indicates that I_o also increases by delaying the turn-on angle close to the aligned position and advancing the turn-off angle. The latter trend does not represent the actual characteristics of the generator but is an error as a result of simplifications and the use of logarithm functions to derive the inverse machine model. The switching angles in those regions will not be used in this particular example.

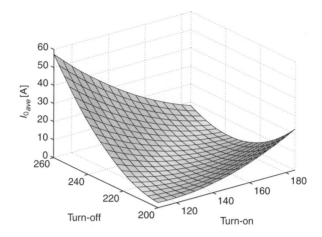

Fig. 10.17 I_o vs. switching angles (angles in electrical degrees).

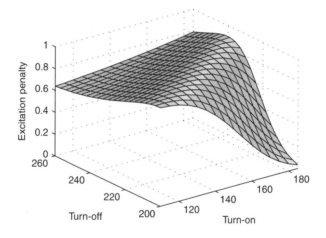

Fig. 10.18 Excitation penalty vs. switching angles (angles in electrical degrees).

controller input and output may be obtained through intensive experimental work, by simulation or analytical method. The most accurate data will be given by experiment but it may be time consuming and not economical. The accuracy of simulations will be influenced by the accuracy of parameters used and obtaining accurate parameters is not an easy task and may require a certain amount of experimental work. The advantage of analytical methods is that the controller can be designed before an actual system is implemented. However, its accuracy should be always carefully examined.

10.6 Implementation of the controller

Figure 10.19 shows an example of the implementation of the generator controller. In the example a microcontroller is used and the dotted part in Figure 10.12 is implemented by software.

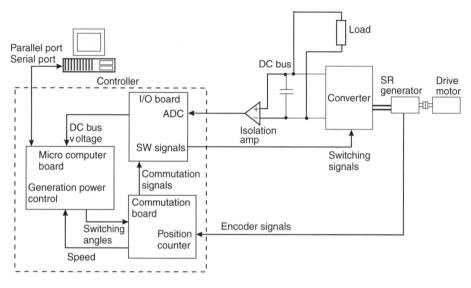

Fig. 10.19 Implementation of the generator controller.

The controller consists of three main parts: microcontroller board, commutation board and I/O board. The serial interface is used as a user interface for setting parameters of the controller. The commutation board includes an FPGA (field programable gate array), encoder signal receiver and D/A converter. The commutator implemented in the FPGA determines the switching states of the converter from the incoming encoder signals and commutation angle information from the microcontroller. The commutation signals are sent to the I/O board. The D/A converter is used for the current regulation and overcurrent detection. The I/O board outputs the commutation signals through optical transmitters to the power converter. The d.c.-bus voltage is fed back through an isolation amplifier and an A/D converter. The voltage information is processed for the generation power control.

10.7 Fault-tolerant operation

The switched reluctance generator has been receiving an increasing attention for aircraft applications where robustness, high speed and fault tolerance are important (Ferreira *et al.*, 1995; MacMinn, 1989; Radun, 1994; Richter and Ferreira, 1995; Richter, 1988). The classic converter topology (Miller, 1993) avoids the problem of shoot-through faults found in a.c. inverter technology. Furthermore, the short-pitched windings exhibit a large amount of electromagnetic independence, which may be exploited by appropriate converter as well as machine topologies.

Much interest has been paid to the possibility of replacing the 400 Hz a.c. power distribution found on many aircraft with a d.c. system at 270 V. The switched reluctance machine has been identified as a potential candidate for the generator in such systems (Radun, 1994).

However, relatively little has been reported on switched reluctance machines operating under fault conditions. Stephens (1991) suggested that switched reluctance motors could operate by disconnecting faulty windings with the use of special fault detectors. Miller (1995) reported an analysis of operation under faults and compared several winding configurations. Switched reluctance generator converter topologies for improved fault-tolerance have been studied by Richter and Ferreira. Radun (1994) suggested a dual-converter topology which has separate excitation and load buses, and MacMinn (1989) studied the design of the power stage for switched reluctance generators.

In this section the fault tolerance of the multi-pole single-phase switched reluctance generator, which has interesting features, is described.

10.7.1   The single-phase generator

The cross-section of the prototype single-phase 8/8-pole machine described in Hayashi and Miller (1995) is shown in Figure 10.20. The machine is wound with one coil per pole and the number of turns per pole is 35. The eight coils were initially connected such that adjacent poles have opposite magnetic polarity, producing so-called short flux paths. This has been reported to result in lower iron losses (Michaelides and Pollock, 1995b), which is of importance especially in high-speed applications. The 8/8 machine has narrower stator pole-arcs than the equivalent 8/6 machine, hence a wider slot area is available, resulting in lower copper losses. On the other hand, a narrower pole-arc may also reduce the aligned inductance and increase the pole flux density. For a more detailed discussion on the choice of pole numbers, see Lovatt and Stephenson (1992). Apart from a high efficiency, the 8/8 machine appears attractive as its coils may be

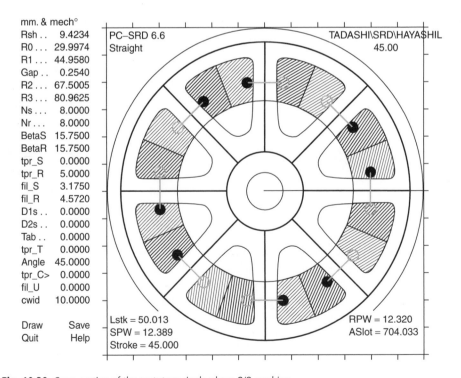

mm. & mech°	
Rsh ..	9.4234
R0 ...	29.9974
R1 ...	44.9580
Gap ..	0.2540
R2 ...	67.5005
R3 ...	80.9625
Ns ...	8.0000
Nr ...	8.0000
BetaS	15.7500
BetaR	15.7500
tpr_S	0.0000
tpr_R	5.0000
fil_S	3.1750
fil_R	4.5720
D1s ..	0.0000
D2s ..	0.0000
Tab ..	0.0000
tpr_T	0.0000
Angle	45.0000
tpr_C>	0.0000
fil_U	0.0000
cwid	10.0000

Draw   Save
Quit   Help

PC–SRD 6.6
Straight

TADASHI\SRD\HAYASHIL
45.00

Lstk = 50.013
SPW = 12.389
Stroke = 45.000

RPW = 12.320
ASlot = 704.033

Fig. 10.20 Cross-section of the prototype single-phase 8/8 machine.

connected to minimize the impact of faults. In the following sections, several winding configurations are studied.

10.7.2 Fault-tolerant converter for the single-phase generator

First, it is shown how the machine can continue to operate with a loss (open-circuit) of one or more coils. To tolerate an open-coil fault, the coils must be connected in parallel (two banks of four coils in parallel or alike, with an increase in converter complexity). The split-winding converter shown in Figure 10.21 was proposed in Sawata *et al.* (1999a) and assumed to be used in the following analysis. One of the two banks of coils can be disconnected from the power source when a fault in coils occurs with this converter. Under normal condition S1, S2 and S3 receive identical switching signals. The excitation current flows from the positive d.c.-bus through S1, coils, S2 and S3 to the negative bus. The generated current flows from the negative bus through D1, coils, D2 and D3 to the positive bus, then charges the capacitor. In order to disconnect coil bank B, S3 is switched off and the generator is operated by controlling S1 and S2. If coil bank B does not share flux paths with bank A, the operation of coil bank B could be shut off completely. However, during the shut-off process, uncontrollable transient current could flow through diode D3.

Other single-phase generator topologies can be connected in a similar fashion, but the converter in Figure 10.21 has the drawback of disconnecting half the total number of coils even when just one coil is faulty. To separate a smaller number of coils, increased circuit complexity is required. Figure 10.22 shows one such example.

There are several ways to configure the magnetic polarity of the coils with the converter in Figure 10.21. For normal operation the 8/8 machine should have four pairs of N and S coils. For a fault operation in which half the coils are disconnected

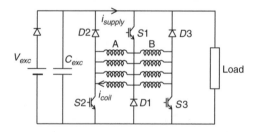

Fig. 10.21 Fault-tolerant single-bus converter circuit for split-winding single-phase generator.

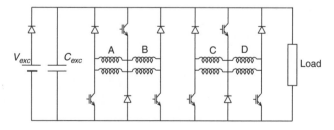

Fig. 10.22 Fault-tolerant single-bus converter circuit for split winding with four banks for the single-phase generator.

Table 10.3 Definition of open-coil faults

Case I	Normal condition	(NSNSNSNS)
Case II	4-coil open	(NONOSOSO)
Case III	4-coil open	(NOSONOSO)
Case IV	4-coil open	(NONONONO)
Case V	1-coil open	

O refers to an open-circuit coil

(open-circuited), the polarity of remaining coils could be either two pairs of N and S or four of either polarity from a symmetrical point of view. Here, the five cases in Table 10.3 are examined. The study of 1-coil open-circuit fault is a good example to show how the magnetic unbalance affects the operation. The 1-coil short-circuit fault is also discussed in the later sections although it is not listed here.

Figures 10.23, 10.24 and 10.25 show the flux paths of the machine when the healthy coils are excited by a coil current of 10 A, calculated using finite-element analysis

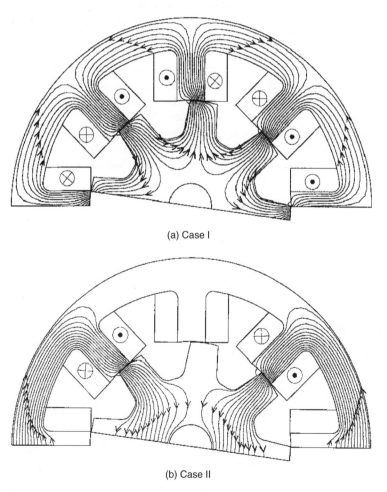

(a) Case I

(b) Case II

Fig. 10.23 Flux-plots calculated by finite-element analysis; Cases I and II.

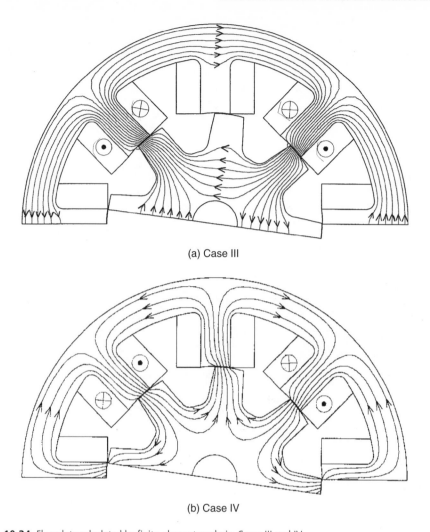

(a) Case III

(b) Case IV

Fig. 10.24 Flux-plots calculated by finite-element analysis; Cases III and IV.

(FEA). Figure 10.23(a) shows the short flux paths of Case I, corresponding to the rated phase current of 80 A with the converter of Figure 10.21. Figures 10.23(b) and 10.24(a) show the flux paths of Case II and III respectively. For Case II and III, the flux in open-circuit coil poles is cancelled by the MMF of the conducting coils under linear conditions. Flux concentration in stator and rotor yoke sections is higher for Case II than for Case III. For Case IV shown in Figure 10.24(b), the flux paths remain unchanged from Case I although the flux density decreases with decreased MMF level. When one coil is unexcited (Case V) the distribution of flux is not symmetrical any more, as shown in Figure 10.25. This may result in unbalanced lateral forces (Miller, 1995) and lead to mechanical failure. Therefore, this fault should be cleared as soon as possible. With the single-phase generator system, the magnetic balance can be maintained and the problem of lateral forces can be avoided by disconnecting a coil bank when one of the coils is faulty.

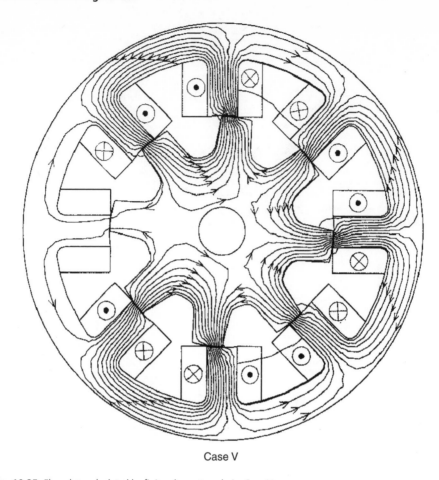

Case V

Fig. 10.25 Flux-plots calculated by finite-element analysis; Case V.

10.7.3   Open-coil fault analysis

The performance of the single-phase switched reluctance machine under faults can be estimated by linear analysis as follows.

The equivalent magnetic circuit of the machine is shown in Figure 10.26. The MMF potential is defined 0 at the stator yoke and F_0 at the rotor yoke. F_k represents the MMF of coil 'k', R_k represents the airgap reluctance and ϕ_k represents the flux through the pole. F_0 and ϕ_k are expressed by the following equation:

$$F_0 = \frac{\displaystyle\sum_k \frac{F_k}{R_k}}{\displaystyle\sum_k \frac{1}{R_k}} \tag{10.18}$$

$$\phi_k = \frac{F_k - F_0}{R_k}. \tag{10.19}$$

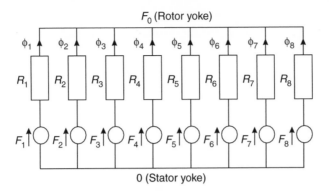

Fig. 10.26 Equivalent magnetic circuit of the single-phase machine.

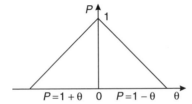

Fig. 10.27 Normalized permeance profile.

Since all airgap reluctances are the same for the single-phase machine, it follows that:

$$F_0 = \frac{\sum_k F_k}{8}.$$ (10.20)

For a simple comparison between normal and faulted conditions, a normalized permeance profile shown in Figure 10.27 is assumed.

Using this permeance profile, torque produced on each pole is expressed as follows:

$$T = \frac{1}{2}(F_k - F_0)^2 \frac{dP}{d\theta} = \frac{1}{2}(F_k - F_0)^2.$$ (10.21)

Case I
The MMF in each pole is such that:

$$F_1 = F_3 = F_5 = F_7 = N_p i$$ (10.22)

$$F_2 = F_4 = F_6 = F_8 = -F_1$$ (10.23)

where N_p is the number of turns per pole. This gives total torque T_I for Case I:

$$T_I = 8T_k = 4N_p^2 i^2.$$ (10.24)

This torque will be compared with the torque produced during faults.

Cases II and III

Cases II and III give the same result with linear analysis. The configuration of the MMF is as follows:

$$F_1 = F_5 = -F_3 = -F_7 = N_p i \tag{10.25}$$

$$F_2 = F_4 = F_6 = F_8 = 0. \tag{10.26}$$

Then the total torque $T_{II,III}$ for Cases II and III is:

$$T_{II,III} = 2N_p^2 i^2 = 0.5T_I. \tag{10.27}$$

Case IV

The MMF in each pole is such that:

$$F_1 = F_3 = F_5 = F_7 = N_p i \tag{10.28}$$

$$F_2 = F_4 = F_6 = F_8 = 0. \tag{10.29}$$

The total torque T_{IV} for Case VI is:

$$T_{IV} = N_p^2 i^2 = 0.25T_I. \tag{10.30}$$

Case V

The MMF of each coil is as follows:

$$F_1 = F_3 = F_5 = F_7 = N_p i \tag{10.31}$$

$$F_4 = F_6 = F_8 = -N_p i \tag{10.32}$$

$$F_2 = 0. \tag{10.33}$$

Then the total torque T_V for Case V is:

$$T_V \approx 3.44N_p^2 i^2 = 0.86T_I. \tag{10.34}$$

Table 10.4 summarizes the analysis. It shows that the single-phase generator could generate half the normal power when four of eight coils are conducting, with winding configuration of either Case II or III.

10.7.4 2-D finite-element analysis and static torque measurement

Measured static torque curves of the prototype single-phase generator are shown in Figure 10.28 for a coil current of 15 A. The static torque of the machine for Case II

Table 10.4 Torque production analysis under open-coil faults

Faults	Torque (vs. Case I)
Cases II, III	50%
Case IV	25%
Case V	86%

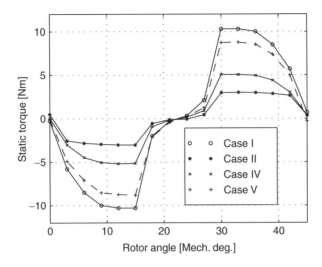

Fig. 10.28 Measured static torque curves.

is 50% of Case I. The same measurement was carried out for Case III and it gave the same result as Case II although this is not shown in Figure 10.28. The static torque for Case IV is 25% of that of Case I. When one of the eight coils is open-circuited (Case V) the static torque is 86% of Case I. When the coils are split into two banks, the magnetic polarity of Case II should be configured. The measurement results support the analytical predictions well.

The result confirms that the available power per coil current during faults is:

$$P_{fault} = P_{normal} \frac{N - n}{N} \qquad (10.35)$$

where N is the number of coils and n the number of faulted coils. The single-phase generator could generate half its normal energy when half the coils are disconnected and currents in the remaining coils are normal.

10.7.5   Coil short-circuit fault

With this fault the d.c.-bus is short-circuited when the switches are on and a large current may flow. In order to prevent the failure of the semiconductor switches and loss of winding insulation, the overcurrent should be detected by fault detectors and the faulty bank should be disconnected as soon as possible.

Apart from the d.c.-bus short-circuit, an induced current in the short-circuited coil could be a problem when the faulty coil is magnetically coupled with the healthy coils. An example of the induced current is shown in Figure 10.29. It shows the measured current in the faulty coil when one of the S-coils is short-circuited and physically disconnected from the converter. The current in S-coils I_S differs very much from the normal one and the induced current in the faulty coil I_{Fault} has a comparable shape and amplitude to I_S. Under this condition the generating operation may not be sustained.

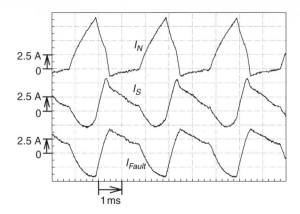

Fig. 10.29 Current waveforms when one coil is short-circuited.

10.7.6   Disconnection of faulty coils

Figure 10.30 shows current waveforms when the faulty bank is disconnected by turning off the semiconductor switch which corresponds to the faulty bank. The currents in the disconnected coils I_{NB} and I_{SB} decrease with time while the current in S-coils in the healthy bank I_{SA} recovers and becomes the same level as that in the N-coils I_{NA}. The circulating currents in the healthy bank decrease but do not become zero. The reason may be that a perfect balance in the healthy coils is not obtained because of the leakage inductance and the partial saturation of the steel in the prototype machine.

Figure 10.31 shows the current waveforms when one coil is short-circuited by a fault and the faulty bank is disconnected by turning off the corresponding lower semiconductor switch. The current in the short-circuited coil is very small since the magnetic coupling between the banks is very small when one bank is turned off.

This result suggests that the proposed converter manages coil short-circuit fault with proper fault detectors. Detailed discussion on the fault detectors for switched reluctance machines is seen in Stephens (1989). This is not discussed here in detail but shows a

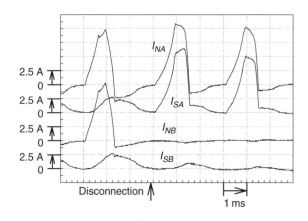

Fig. 10.30 Current waveforms when disconnecting the faulty bank.

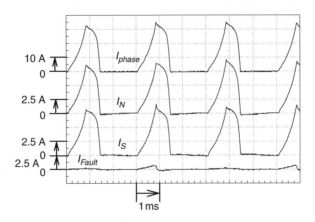

Fig. 10.31 One coil short-circuit fault.

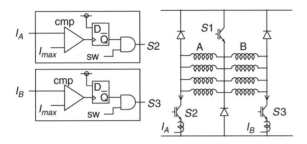

Fig. 10.32 Example of fault detector.

simple example for the split-winding converter. An example of simple implementations of the detector is shown in Figure 10.32. One current sensor may be used for each leg to detect overcurrents. When the current in a leg exceeds the threshold I_{max} the fault is detected and the flip-flop holds the signal. The output of the flip-flop blocks the switching signal SW to the semiconductor switch in the faulty leg. It is an advantage of the proposed converter that only two current detectors are required.

Bibliography

Acarnley, P.P., *Stepping motors: a guide to modern theory and practice*, Peter Peregrinus Ltd, 1982.

Acarnley, P.P., Hill, R.J. and Hooper, C.W., Detection of rotor position in stepping and switched reluctance motors by monitoring of current waveforms, *IEEE Transactions*, Vol. IE-32, No. 3, pp. 215–222, August 1985.

Acarnley, P.P., French, C.D. and Al-Bahadly, I.H., Position estimation in switched-reluctance drives, *Proceedings of European Power Electronics Conference*, pp. 3.765–3.770, 1995.

Alexanderson, E.F.W., *Apparatus for producing electromotive force of special waveform*, US Patent No. 1,250,752, 18 December 1917.

Amor, L.B., Dessaint, L.-A., Akhrif, O. and Olivier, G., Adaptive input-output linearization of a switched reluctance motor for torque control, *IEEE-IECON Conf. Record*, Maui, pp. 2155–2160, 1993a.

Amor, L.B., Dessaint, L.-A., Akhrif, O. and Olivier, G., Adaptive feedback linearization for position control of a switched reluctance motor: analysis and simulation, *International Journal of Adaptive Control and Signal Processing*, Vol. 7, pp. 117–136, 1993b.

Amor, L.B., Dessaint, L.-A. and Akhrif, O., Adaptive nonlinear torque control of a switched reluctance motor via flux observation, *Mathematics and Computers in Simulation*, Vol. 38, pp. 345–358, 1995.

Anderson, A.F., *The thyristor control of a reluctance motor of improved rotor construction*, Ph.D. Thesis, University of St Andrews, Queens College Dundee, 1966.

Anderson, A.F., William Henley, pioneer instrument maker and cable manufacturer 1813 to 1882, *IEE Proceedings*, Vol. 132, Pt A, No. 4, 249–261, July 1985.

Arkadan, A.A., Shehadeh, H.H., Brown, R.H. and Demerdash, N.A.O., Effects of chopping on core losses and inductance profiles of SRM drives, *IEEE Transactions on Magnetics*, Vol. 33, No. 2, pp. 2105–2108, March 1997.

Austermann, R., *Circuit arrangement for commutating a reluctance motor*, US Patent No. 5180960, 19 January 1993.

Bahn, I., *Three-phase reluctance motor*, EPA 0 444 203, 18 September 1989.

Barnes, M. and Pollock, C., Power converter for single phase switched reluctance motors, *IEE Electronics Letters*, Vol. 31, No. 25, pp. 2137–2138, 7 December 1995.

Barnes, M. and Pollock, C., New class of dual voltage converters for switched reluctance drives, *IEE Proceedings in Electric Power Applications*, Vol. 145, No. 3, pp. 164–168, May 1998.

Barnes, M. and Pollock, C., Power electronic converters for switched reluctance drives, *IEEE Transactions on Power Electronics*, Vol. 13, No. 6, November 1998.

Barrass, P.G. and Mecrow, B.C., *Flux and torque control of switched reluctance machines*, IEE Proc.-Electr. Power Appl., Vol. 145, No. 6, pp. 519–527, November 1998.

Bartos, R.P., Houle, T.H. and Johnson, J.H., *Switched reluctance motor with sensorless position detection*, US Patent No. 5256923, 26 October 1993.

Bass, J.T., Ehsani, M. and Miller, T.J.E., Robust torque control of switched-reluctance motors without a shaft position sensor, *IEEE Transactions on Industrial Electronics*, Vol. IE-33, pp. 212–216, 1986.

Bass, J.T., Miller, T.J.E. and Steigerwald, R.L., Development of a unipolar converter for variable reluctance motor drives, *IEEE Transactions on Industry Applications*, Vol. IA-23, pp. 545–553, 1987.

Bausch, H. and Rieke, B., Speed and torque control of thyristor-fed reluctance motors, *Proc. Int. Conf. Electric Machines*, Vienna, Pt I, pp. 128.1–128.10, 1978.

Bausch, H. and Rieke, B., Performance of thyristor-fed electric car reluctance machines, *Proc. Int. Conf. Electric Machines*, Brussels pp. E4/2.1–2.10, 1980.

Bedford, B.D., *Compatible permanent magnet or reluctance brushless motors and controlled switch circuits*, US Patent No. 3,678,352, 18 July 1972a.

Bedford, B.D., *Compatible brushless reluctance motors and controlled switch circuits*, US Patent No. 3,679,953, 25 July 1972b.

Besbes, M., Picod, C., Camus, F. and Gabsi, M., Influence of stator geometry on vibratory behaviour and electromagnetic performances of switched reluctance motors, *IEE Proc.-Electr. Power Appl.*, Vol. 145, No. 5, pp. 462–468, September 1998.

Betz, R.E., Lagerquist, R., Jovanovic, M., Miller, T.J.E. and Middleton, R.H., Control of synchronous reluctance machines, *IEEE Transactions on Industry Applications*, Vol. 29, No. 6, pp. 1110–1122, 1993.

Bianco, C.G.L. *et al*; A prototype controller for variable reluctance motors, *IEEE Transactions on Industrial Electronics*, Vol. 43, No. 1, pp. 207–216, February 1996.

Bin-Yen, Ma Tian-Hua, Liu Ching-Guo, Chen Tsen-Jui and Shen Wu-Shiung Feng, Design and implementation of a sensorless switched reluctance drive system, *Proceedings of PEDS'97 Conference*, Singapore, pp. 174–180, 1997.

Blaabjerg, F., Christensen, L., Hansen, S., Kristofferson, J.P. and Rasmussen, P.O., Sensorless control of a switched reluctance motor with variable-structure observer, *Electromotion*, pp. 141–152, 1996.

Blaabjerg, F., Kjaer, P.C., Rasmussen, P.O. and Cossar, C., Improved digital current control methods in switched reluctance motor drives, *IEEE Transactions*, Vol. 14, No. 3, pp. 563–572, 1999.

Blackburn, S.E., *Apparatus and method for generating rotor position signals and controlling commutation in a variable reluctance motor*, US Patent No. 5783916, 21 July 1998.

Boldea, I., *Reluctance Synchronous Machines and Drives*, Oxford Science Publications, Clarendon Press, Oxford, 1996.

Bose, B.K., Miller, T.J.E., Szczesny, P.M. and Bicknell, W.H., Microcomputer control of switched reluctance motors, *IEEE Transactions on Industry Applications*, Vol. 22, No. 4, pp. 253–265, 1985.

Bose, B.K., *Power Electronics & AC Drives*, Prentice Hall, 1986.

Bose, B.K. (ed.), *Microcomputer Control of Power Electronics and Drives*, IEEE Press, 1987.

Bose, B.K., *Control system for switched reluctance motor*, US Patent No. 4,707,650, 17 November 1987.

Bowers, B., *Sir Charles Wheatstone*, Science Museum, pp. 72–85.

Buerhing and Freris, L.L., Control policies for wind-energy conversion systems, *IEE Proceedings*, Part C, Vol. 128, No. 5, pp. 253–261, September 1981.

Byrne, J.V., DT 20 30789 B2 (German patent) 23 June 1970 (controlled saturation); also Byrne, J.V. and Lacy, J.G. *Electrodynamic system comprising a variable reluctance machine*, UK Patent 1,321,110, 24 June 1970.

Byrne, J.V. and Lacy, J.C., *Electrodynamic system comprising a variable reluctance machine*, US Patent No. 3,956,678, 11 May 1976a.

Byrne, J.V. and Lacy, J.G., Characteristics of saturable stepped and reluctance motors, IEE Conference Publication No. 136, *Small Electrical Machines*, pp. 93–6, 1976b.

Byrne, J.V. and McMullin, M.F., Design of a reluctance motor as a 10kW spindle motor, *Proceedings of the PCI/Motor-Con Conference*, Geneva, pp. 10–24, September 1982.

Byrne, J.V., McMullin, M.F. and O'Dwyer, J.B., A high performance variable reluctance drive: a new brushless servo, *Motor-Con Proceedings*, pp. 147–160, October 1985.

Cailleux, H., Le Pioufle, B., Multon, B. and Sol, C., A precise analysis of the phase commutation for the torque nonlinear control of a switched reluctance motor – torque ripples minimization, *IEEE-IECON Conf. Record*, Maui, pp. 1985–1990, 1993.

Cameron, D.E., Lang, J.H. and Umans, S.D., The origin and reduction of acoustic noise in doubly salient variable-reluctance machines, *IEEE Transactions on Industry Applications*, Vol. IA-26, No. 6, pp. 1250–1255, 1992.

Cameron, D.E. and Lang, J.H., The control of high-speed variable-reluctance generators in electric power systems, *IEEE Transactions on Industry Applications*, Vol. 29, No. 6, November/December 1993.

Camus, F., Gabsi, M. and Multon, B., Prediction des vibrations du stator d'une machine a reluctance variable en fonction du courant absorbe, *J. Phys. III France 7* (1997) 387–404, February 1997.

Chan, C.C., Single-phase switched reluctance motors, *IEE Proceedings*, Vol. 134, Pt B, No. 1, pp. 53–56, January 1987.

Chappell, P.H., Ray, W.F. and Blake, R.J., Microprocessor control of a variable-reluctance motor, *IEE Proceedings*, Vol. 131, Pt B, No. 2, pp. 51–60, March 1984.

Chappell, P.H., Current pulses in switched reluctance motor, *IEE Proceedings*, Vol. 135, Pt B, No. 5, pp. 224–230, September 1988.

Charlisha, An idea 100 years old comes to fruition, *Financial Times*, 26 January, 1983.

Cheok, A. and Ertugrul, N., A model free fuzzy logic based rotor position sensorless switched reluctance motor drive, *Conf. Rec. IEEE Industry Applications Society Annual Meeting*, pp. 76–83, 1996.

Chiba, A. and Fukao, T., A closed-loop operation of super high-speed reluctance motor for quick torque response, *IEEE Transactions on Industry Applications*, Vol. 28, No. 3, pp. 600–606, 1992.

Chuang, T.A. and Pollock, C., Robust speed control of a switched reluctance vector drive using variable structure approach, *IEEE Transactions on Industrial Electronics*, Vol. 44, No. 6, December 1997.

Colby, R.S., Mottier, F. and Miller, T.J.E., Vibration modes and acoustic noise in a 4-phase switched reluctance motor, *IEEE Transactions on Industry Applications*, Vol. 32, No. 6, pp. 1357–1364, November–December 1996.

Compter, J.C., Microprocessor controlled single-phase reluctance motor, *Drives/Motors/Controls Conference*, Brighton, UK, 64–68, 24–26 October 1984.

Compter, J.C., *Single-phase reluctance motor*, EP 0 163 328 B, 18 April 1985.

Corda, J. and Stephenson, J.M., Analytical estimation of the minimum and maximum inductances of a double-salient motor, *Leeds Int. Conf. On Stepping Motors and Systems*, Leeds, pp. 50–59, September 1979.

Corda, J. and Stephenson, J.M., Speed control of switched reluctance machines, *ICEM-82*, Budapest, pp. 235–238, 1982.

Corda, J., Masic, S. and Stephenson, J.M., Computation and experimental determination of running torque waveforms in switched reluctance motors, *IEE Proceedings*, Pt B, Vol. 140, No. 6, pp. 387–392, 1993.

Cossar, C. and Miller, T.J.E., Electromagnetic testing of switched reluctance motors, *International Conference on Electrical Machines (ICEM)*, Manchester, pp. 470–474, 15–17 September 1992.

Cruickshank, A.J.O. and Anderson, A.F., Development of the reluctance motor, *Electronics and Power*, Vol. 12, 48, 1966.

Cruickshank, A.J.O., Menzies, R.W. and Anderson, A.F., *IEE Proceedings*, Vol. 113, No. 12, pp. 2058–2060, December 1966b.

Cruickshank, A.J.O. and Anderson, A.F., UK Patent No. 1,114,561, 1968.

Cruickshank, A.J.O., Anderson, A.F. and Menzies, R.W., Theory and performance of reluctance motors with axially laminated anisotropic rotors, *IEE Proceedings*, Vol. 118, No. 7, pp. 887–894, July 1971.

Cruickshank, A.J.O., Anderson, A.F. and Menzies, R.W., *Stability of reluctance motors from freely accelerating torque speed curves*, Paper No T 72 049–0, IEEE Winter Meeting, New York, NY, 30 January–4 February 1972.

Davis, R.M., A variable-speed drive that took a century and a half to develop, *Electrical Times*, pp. 6–9, 27 May, 1983.

DiRenzo, M.T. and Khan, W., Self-trained commutation algorithm for an SR motor drive system without position sensing, *Conf. Rec. IEEE Industry Applications Society Annual Meeting*, pp. 341–348, October 1997.

Dissauer, G. and Jones, D., Design secrets of switched-reluctance motors, *Machine Design*, 26 September 1996.

Dorf, R.C., *Modern Control Systems*, Addison-Wesley, ISBN 0–201-01258–8, 1980.

Dhyanchand, P., *Power conversion system using a switched reluctance motor/generator*, US Patent No. 5,012,177, 30 April 1991.

Drager, B.T., Jones, S.R. and Fordyce, G.T., *Relative angle estimation apparatus for a sensorless switched reluctance machine system*, US Patent No. 5867004, 2 February 1999.

Egan, M.G., Murphy, J.M.D., Kenneally, P.F., Lawton, J.V. and McMullin, M.F., A high performance variable reluctance drive: achieving servomotor control, *MOTORCON* (Chicago), pp. 161–168, October 1985.

Egan, M.G., Harrington, M.B. and Murphy, J.M.D., PWM-based position sensorless control of variable reluctance motor drives, *Proceedings of European Power Electronics Conference*, Firenze, pp. 4-024–4-029, 1991.

Ehsani, M., *Position sensor elimination technique for the switched reluctance motor drive*, US Patent No. 5072166, 10 December 1991.

Ehsani, M., Husain, I. and Kulkarni, A.B., Elimination of discrete position sensor and current sensor in switched reluctance motor drives, *IEEE Transactions on Industry Applications*, Vol. IA-28, No. 1, pp. 128–135, January–February 1992.

Ehsani, M., Husain, I., Mahajan, S. and Ramani, K.R., New modulation encoding techniques for indirect rotor position sensing in switched reluctance motors, *IEEE Transactions on Industry Applications*, Vol. IA-30, No. 1, pp. 85–91, 1994.

Ehsani, M., *Method and apparatus for sensing the rotor position of a switched reluctance motor without a shaft position sensor*, US Patent No. 5291115, 1 March 1994.

Ehsani, M., *Method and apparatus for sensing the rotor position of a switched reluctance motor*, US Patent No. 5410235, 25 April 1995.

Elmas, C. and Zelaya-De la Parra, H., Position sensorless operation of a switched reluctance drive based on observer, *Proceedings of European Power Electronics Conference*, pp. 82–87, 1993.

Ertugrul, N. and Acarnley, P.P., A new algorithm for sensorless operation of permanent magnet motors, *IEEE Transactions on Industry Applications*, Vol. 30, No. 1, pp. 126–133, 1994.

Favre, E., Cardoletti, L. and Jufer, M., Permanent-magnet synchronous motors: a comprehensive approach to cogging torque suppression, *IEEE Transactions on Industry Applications*, Vol. 29, No. 6, pp. 1141–1149, 1993.

Ferreira, C.A., Jones, S.R., Heglund, W.S. and Jones, W.D., Detailed design of a 30-kW switched reluctance starter/generator system for a gas turbine engine application, *IEEE Transactions on Industry Applications*, Vol. 31, No. 3, pp. 553–561, May/June 1995.

Filicori, F., Guarino Lo Bianco, C. and Tonielli, A., Modeling and control strategies for a variable reluctance direct-drive motor, *IEEE Transactions on Industrial Electronics*, Vol. 40, No. 1, pp. 105–115, 1993.

Finch, J.W., Harris, M.R., Musoke, A. and Metwally, H.M.B., Variable-speed drives using multi-tooth per pole switched reluctance motors, *13th Incremental Motion Controls Symposium*, University of Illinois, Urbana-Champaign, Illinois, USA, pp. 293–302, 1984.

Finch, J.W., Metwally, H.M.B. and Harris, M.R., Switched reluctance motor excitation current: scope for improvement, *Proceedings of Power Electronics and Variable-Speed Drives Conference*, London, IEE Conference Publication No. 264, pp. 196–199, 1986.

Fisch, J.H., Li, Y., Kjaer, P.C., Gribble, J.J. and Miller, T.J.E., Pareto-Optimal firing angles for switched reluctance motor control, *Proceedings of 2nd International Conference on Genetic Algorithms in Engineering Systems: Innovations and Applications*, University of Strathclyde, Glasgow, UK, pp. 90–96, 2–4, September 1997.

Flower, J.O., Richardson, K.M. and Pollock, C., An experimental integrated switched reluctance propulsion unit: design, construction and preliminary results, *Transactions of the Institute of Marine Engineers*, Vol. 108, Part 2, pp. 127–140, 1996.

French, P. and Williams, A.H., (TRW Inc., Cleveland Ohio), *A new electric propulsion motor*, AIAA 3rd Propulsion Joint Specialist Conference; AIAA paper No. 67 523, Washington DC, 17–21 July, 1967.

Fulton, N.N. and Randall, S.P., Switched reluctance generators for wind energy applications, *Proceedings of the 8th British Wind Energy Association*, pp. 211–218, 1986.

Furmanek, R., French, A. and Horst, G.E., Horizontal axis washers, *Appliance Manufacturer*, 52–53, March 1997.

Gallegos-Lopez, G., *Sensorless control for switched reluctance motor drives*, PhD thesis, University of Glasgow, Glasgow, UK, 1998.

Gallégos-Lopez, G., Kjaer, P.C. and Miller, T.J.E., A new sensorless method for switched reluctance motor drives, *IEEE Transactions on Industry Applications*, Vol. 34, No. 4, pp. 832–840, July–August 1998.

Gallegos-Lopez, G., Kjaer, P.C. and Miller, T.J.E., High-grade position estimation for SRM drives using flux-linkage/current correction model, *IEEE Transactions on Industry Applications*, Vol. 35, No. 4, pp. 859–869, July/August 1999.

Goldenberg, A.A., Laniado, I., Kuzan, P. and Zhou, C., Control of switched reluctance motor torque for force control applications, *IEEE Transactions on Industrial Electronics*, Vol. 41, No. 4, pp. 461–466, 1994.

Gribble, J.J., Kjaer, P.C., Cossar, C. and Miller, T.J.E., Optimum commutation angles for current-controlled switched reluctance motors, *Proceedings of Power Electronics and Variable Speed Drives Conference*, Nottingham, pp. 87–92, 1996.

Gribble, J.J., Kjaer, P.C. and Miller, T.J.E., Optimal commutation in average torque control of switched reluctance motors, *IEE Proc.-Elect. Power Appl.*, Vol. 146, No. 1, pp. 2–10, January 1999.

Hancock, C.J. and Hendershot, J.R., *Electronically commutated reluctance motor*, PCT International patent application No. W.O. 90/11641, 4 October 1990.

Harris, M.R., Andjargholi, V., Hughes, A., Lawrence, P.J. and Etran, B., Limitations on reluctance torque in doubly-salient structures, *Int. Conf. On Stepping Motors and Systems*, Leeds, pp. 158–168, 15–18 July 1974.

Harris, M.R., Finch, J.W., Mallick, J.A. and Miller, T.J.E., A review of the integral-horsepower switched reluctance drive, *IEEE Transactions on Industry Applications*, Vol. IA-22, No. 4, 716–72, 15–18 July 1986.

Harris, W.D. and Lang, J.H., A simple motion estimator for VR motors, *IEEE Transactions on Industry Applications*, Vol. IA-26, No. 2, pp. 237–243, March/April 1990.

Harris, W.A., Goetz, R. and Stalsberg, K.J., *Switched reluctance motor position by resonant signal injection*, US Patent No. 5,196,775, 23 March 1993.

Hayashi, Y. and Miller, T.J.E., Single-phase multi-pole switched reluctance motor for solar-powered vehicle, *Proceedings of International Power Electronics Conference*, Yokohama, pp. 575–579, 1995.

Hedlund, G., *D.C. motor with multi-tooth poles*, US Patent No. 4,75,48,362, 31 May 1988.

Hedlund, G. and Lundberg, H., *Motor energizing circuit*, US Patent No. 4,868,478, 19 September 1989.

Hedlund, G., *A device for controlling a reluctance motor*, International Patent No. W09016111, 27 December 1990.

Hedlund, G. and Lundberg, H., *Energizing system for a variable reluctance motor*, US Patent No. 5,043,643, 27 August 1991.

Hendershot, J.R., Short flux paths cool SR motors, *Machine Design*, 106–111, 21 September 1989.

Hendershot, J.R., A comparison of AC, brushless and switched reluctance motors, *Motion Control*, pp. 16–20, 16 April 1991.

Hendershot, J.R., *Polyphase switched reluctance motor*, US Patent No. 5,111,095, 5 May 1992.

Hill, R.J. and Acarnley, P.P., *Stepping motors and drive circuits therefor*, US Patent No. 4,520,302, 28 May 1985.

Holling, G., Yeck, M.M. and Brewer, A.J., *Sensorless commutation position detection for brushless motors*, US Patent No. 5,600,218, 4 February 1997.

Holling, G., Yeck, M.M. and Schmitt, M., *Commutation position detection system and method*, US Patent No. 5,821,713, 13 October 1998.

Holtz, J. and Springob, L., Identification and compensation of torque ripple in high-precision permanent magnet motor drives, *IEEE Transactions on Industrial Electronics*, Vol. 43, No. 2, pp. 309–320, 1996.

Honsinger, V.B., Steady-state performance of reluctance machines, *IEEE Transactions on Power Apparatus and Systems*, Vol. PAS-90, No. 1, January/February/1971.

Hopkinson, J., Dynamo electric machinery. *Phil. Trans. Royal Soc.*, Part 1, 1886. Also *Collected Papers*, Cambridge University Press, 190, Vol. 1, pp. 84–121.

Horst, G.E., *Hybrid single-phase variable reluctance motor*, US Patent No. 5,122,697, 16 June 1992. Also European Patent Application 0 455 578 A2, 6 November 1991.

Horst, G.E., *Isolated segmental switch reluctance motor*, US Patent No. 5,111,096, 5 May 1992.

Horst, G.E., *Shifted pole single phase variable reluctance motor*, US Patent No. 5,294,856, 15 March 1994.

Horst, G.E., *Current decay control in switched reluctance motors*, US Patent No. 5446359, August 1995.

Horst, G.E., *Noise reduction in a switched reluctance motor by current profile manipulation*, US Patent No. 5461295, 1995.

Horst, G.E., *Rotor position sensing in a dynamoelectric machine using coupling between machine coils*, US Patent No. 5701064, 23 December 1997.

Hung, J.U., Efficient torque ripple minimization for variable reluctance motors, *IEEE-IECON Conf. Record*, Maui, pp. 1985–1990, 1993.

Husain, I. and Ehsani, M., Torque ripple minimization in switched reluctance motor drives by PWM current control, *Proceedings of Applied Power Electronics Conference*, Orlando, pp. 72–77, 1994a.

Husain, I. and Ehsani, M., Rotor position sensing in switched reluctance motor drives by measuring mutually induced voltages, *IEEE Transactions on Industry Applications*, Vol. IA-30, No. 3, pp. 665–672, 1994b.

Husain, I., Sodhi, S. and Ehsani, M., A sliding mode observer based controller for switched reluctance motor drives, *Conf. Rec. IEEE Industry Applications Society Annual Meeting*, pp. 635–643, 1994.

Hutton, A.J. and Miller, T.J.E., Use of flux screens in switched reluctance motors, *IEE 4th Intl. Conf. On Electrical Machines and Drives*, London, pp. 312–316, September 1989.

Ilic-Spong, M., Miller, T.J.E., MacMinn, S.R. and Thorp, J.S., Instantaneous torque control of electric motor drives, *IEEE Transactions on Power Electronics*, Vol. 2, No. 1, pp. 55–61, 1987a.

Ilic-Spong, M., Marino, R., Peresada, S.M. and Taylor, D.G., Feedback linearizing control of switched reluctance motors, *IEEE Transactions on Automatic Control*, Vol. AC-32, No. 5, pp. 371–379, May 1987b.

Irfan Ahmed, *Digital Control Applications with the TMS320 Family*, Texas Instruments, 1991.

Jones, S.R. and Drager, B.T., Performance of a high-speed switched reluctance starter/generator system using electronic position sensing, *Conf. Rec. IEEE Industry Applications Society Annual Meeting*, pp. 249–253, 1995.

Jones, S.R. and Drager, B.T., *Estimation initialization circuit and method for a sensorless switched reluctance machine system*, US Patent No. 5689165, 18 November 1997.

Jones, S.R. and Drager, B.T., *Absolute angle estimation apparatus for a sensorless switched reluctance machine system*, US Patent No. 5844385, 1 December 1998.

Jones, S.R. and Drager, B.T., *Instantaneous position indicating apparatus for a sensorless switched reluctance machine system*, US Patent No. 5920175, 6 July 1999.

Kalpathi, R.R., *Active phase coil inductance sensing*, US Patent No. 5786681, 28 July, 1998.

Kelly, L., Cossar, C., Miller, T.J.E. and McGilp, M., The SPEED Laboratory Flexible Controller: A Tool for Motor/Drive Performance Evaluation and Optimised Product Development, *Drives & Controls Conference Proceedings*, Telford, UK, March 1999.

Kjaer, P.C., Cossar, C., Gribble, J.J., Li, Y. and Miller, T.J.E., Switched reluctance generator control using an inverse machine model, *Proceedings of International Conference of Electrical Machines*, Paris, pp. 380–385, 1994a.

Kjaer, P.C., Blaabjerg, F., Pedersen, J.K., Nielsen, P. and Andersen, P., A new indirect rotor position detection method for switched reluctance drives, *Proceedings International Conference on Electrical Machines*, Paris, pp. 555–560, 1994b.

Kjaer, P.C., Cossar, C., Gribble, J.J., Li, Y. and Miller, T.J.E., Minimisation of reactive power flow in switched reluctance generator systems, *Proceedings of International Power Electronics Conference*, Yokohama, pp. 1022–1027, 1995.

Kjaer, P.C., Nielson, P., Andersen, L. and Blaabjerg, F., A new energy optimizing control strategy for switched reluctance motors, *IEEE Transactions on Industry Applications*, Vol. 31, No. 5, pp. 1088–1095, 1995.

Kjaer, P.C., Blaabjerg, F., Cossar, C. and Miller, T.J.E., Efficiency optimisation in current controlled variable-speed switched reluctance motor drives, *Proceedings of European Conference on Power Electronics and Applications*, Sevilla, pp. 3.741–3.747, 1995.

Kjaer, P.C., Gribble, J.J. and Miller, T.J.E., Very high bandwidth digital current controller for high performance motor drives, *Proceedings of IEE International Conference on Power Electronics and Variable Speed Drives*, Nottingham, pp. 185–190, 1996.

Kjaer, P.C., Gribble, J.J. and Miller, T.J.E., High grade control of switched reluctance machines, *IEEE Transactions on Industry Applications*, Vol. 33, No. 6, pp. 1585–1593, 1997.

Kjaer, P.C., *High performance control of switched reluctance motors*, Ph.D. Thesis, University of Glasgow, 1997a.

Kjaer, P.C., An alternative method to servo motor drive torque ripple assessment, *European Conference on Power Electronics and Applications*, Trondheim, 1997b.

Kjaer, P.C. and Gallegos-Lopez, G., Single-sensor current regulation in switched reluctance motor drives, *IEEE Transactions on Industry Applications*, Vol. 34, No. 3, pp. 444–451, May/June 1998.

Konecny, K., Analysis of variable reluctance motor parameters through magnetic field simulations, *Motor-Con Proceedings*, p. 2A, 1981.

Konecny, K., European Patent Application EP 0 205 027 B1, 26 May 1986.

Krishnan, R. and Materu, P., Design of a single-switch-per-phase converter for switched reluctance motor drives, *Conf. Rec. IECON-88*, pp. 773–779, 1988.

Krishnan, R., Aravind, S., Bharadwa, J. and Materu, P.N., Computer-aided design of electrical machines for variable speed applications, *IEEE Trans. Ind. Electron.*, Vol. 35, No. 4, pp. 560–571.

Krishnan, R., Arumugam, R. and Lindsay, J.F., Design procedure for switched reluctance motors, *Transactions IEEE*, Vol. IA-24, No. 3, pp. 456–461, May 1988.

Krishnan, R., Mang, X. and Bharadwaj, A.S., Design and performance of a microcontroller based switched reluctance motor drive system, *Electric Machines and Power Systems*, Vol. 18, pp. 359–373, 1991.

Laurent, P., Gabsi, M. and Multon, B., Sensorless rotor position analysis using resonant method for switched reluctance motor, *Conf. Rec. IEEE Industry Applications Society Annual Meeting*, Toronto, pp. 687–694, 2–8 October 1993a.

Laurent, P., Multon, B., Hoang, E. and Gabsi, M., Sensorless rotor position measurement based on PWM eddy current variation for switched reluctance motor, *Proceedings of European Power Electronics Conference*, pp. 3.787–3.792, 1995.

Lawrenson, P.J. and Agu, L.A., A new unexcited synchronous machine, *IEE Proceedings*, Vol. 110, No. 7, July 1963.

Lawrenson, P.J. and Agu, L.A., Theory and performance of polyphase reluctance machines, *IEE Proceedings*, Vol. 111, No. 8, August 1964.

Lawrenson, P.J., Stephenson, J.M., Blenkinsop, P.T., Corda, J. and Fulton, N.N., Variable switched reluctance motors, *Proc. IEE*, Vol. 127, Pt B, No. 4, pp. 253–265, 1980.

Lawrenson, P.J., Switched reluctance drives – a fast growing technology, *Electric Drives and Controls*, pp. 18–23, April/May 1985.

Li, Y. and Tang, Y., Switched reluctance motor drives with fractionally-pitched winding design, *IEEE IAS Conference*, 1998.

Liao, Y., Liang, F. and Lipo, T.A., A novel permanent magnet motor with /doubly salient structure, *IEEE Transactions on Industry Applications*, Vol. 31, No. 5, September/October 1995.

Li, H.-Y., Liang, F., Zhao, Y. and Lipo, T.A., A doubly salient doubly excited variable reluctance motor, *Conf. Rec. IEEE Industry Applications Society Annual Meeting*, Toronto, pp. 137–143, 2–8 October 1993.

Lim, J.Y., *Sensorless switched reluctance motor*, US Patent No. 5,589,751, 31 December 1996.

Lindsay, J.F., Arumugam, R. and Krishnan, R., Finite-element analysis characterization of a switched reluctance motor with multitooth per stator pole, *IEE Proceedings*, Vol. 133, Pt B, pp. 347–353, November 1986.

Lipo, T.A. and Liang, F., *Variable reluctance drive system*, PCT WO 93/23918, 10 May 1993 (d.c. bias winding).

Lipo, T. and Liao, Y., *Variable reluctance motors with permanent magnet excitation*, US Patent No. 5,304,882, 19 April 1994.

Lipo, T. and Liang, D.S., *Transactions IEEE*, Vol. IA-31, pp. 99–106, January–February 1995.

Lipo, T. and Liao, Y., *Transactions IEEE*, Vol. IA-31, pp. 1069–1078, September–October 1995.

Lovatt, H.C. and Stephenson, J.M., Influence of number of poles per phase in switched reluctance motors, *IEE Proceedings*, Part B, Vol. 139, No. 4, pp. 307–314, July 1992.

Lovatt, H.C. and Stephenson, J.M., Computer-optimized current waveforms for switched reluctance motors, *IEE Proc.-Electr. Power Appl.*, Vol. 141, No. 2, pp. 45–51, 1994.

Lumsdaine, A. and Lang, J.H., State observers for variable-reluctance motors, *IEEE Trans.* IE-37, No. 2, pp. 133–142, April 1990.

Lyons, J.P., MacMinn, S.R. and Preston, M.A., Flux/Current methods for SRM rotor position estimation, *Conf. Rec. IEEE Industry Applications Society Annual Meeting*, pp. 482–487, 1991.

Lyons, J.P. and MacMinn, S.R., *Rotor position estimator for a switched reluctance machine*, US Patent No. 5,097,190, 17 March 1992.

Lyons, J.P., MacMinn, S.R. and Preston, M.A., *Rotor position estimator for a switched reluctance machine using a lumped parameter flux/current model*, US Patent No. 5,107,195, 21 April 1992a.

Lyons, J.P. and MacMinn, S.R., *Lock detector for switched reluctance machine rotor position estimator*, US Patent No. 5140244, 18 August 1992b.

Lyons, J.P., MacMinn, S.R. and Preston, M.A., *Discrete position estimator for a switched reluctance machine using a flux-current map comparator*, US Patent No. 5140243, 18 August 1992.

Lyons, J.P. and Preston, M.A., *Low speed position estimator for switched reluctance machine using flux/current model*, US Patent No. 5525886, 11 June 1996.

MacMinn, S.R., Control a switched reluctance aircraft engine starter-generator over a very wide speed range, *Proceedings IECEC*, pp. 631–638, August 1981.

MacMinn, S.R. and Jones, W.D., A very high speed switched reluctance starter-generator for aircraft engine applications, *Proceedings of NAECON-89*, Dayton, OH, 22–26 May 1989.

MacMinn, S.R., Rzesos, W.J., Szczesny, P.M. and Jahns, T.M., Application of sensor integration techniques to switched reluctance motor drives, *Conf. Rec. IEEE Industry Applications Society Annual Meeting*, pp. 584–588, 1988.

MacMinn, S.R. and Roemer, P.B., *Rotor position estimator for switched reluctance motor*, US Patent No. 4772839, 20 September 1988.

MacMinn, S.R., Stephens, C.M. and Szczesny, P.M., *Switched reluctance motor drive system and laundering apparatus employing same*, US Patent No. 4959596, 25 September 1990.

Marcinkiewicz, J.G., Thorn, J.S. and Skinner, J.L., *Improved sensorless commutation controller for a poly-phase dynamoelectric machine*, European Patent Application, 95630049.5, 29 November 1995.

Materu, P. and Krishnan, R., Estimation of switched reluctance motor losses, *Conf. Rec. IEEE Industry Applications Society Annual Meeting*, Pittsburgh, pp. 79–90, 2–7 October 1988.

McCann, R.A., *Method and apparatus for hybrid direct-indirect control of a switched reluctance motor*, US Patent No. 5637974, 10 June 1997.

McCann, R.A. and Husain, I., Application of a sliding mode observer for switched reluctance motor drives, *Conf. Rec. IEEE Industry Applications Society Annual Meeting*, pp. 525–532, 1997a.

McCann, R.A., *Switched reluctance motor with indirect position sensing and magnetic brake*, US Patent No. 5691591, 25 November 1997b.

McCann, R.A., *Switched reluctance motor with indirect position sensing and magnetic brake*, US Patent No. 5949211, 7 September 1999.

McLaughlin, P.J., *Nicholas Callan Priest Scientist 1799–1864*, published by Clonmore & Reynolds and Burns Oates, 1965.

McHugh, P.M., *Current shaping in reluctance machines*, European Patent Application, EP 0 801 464 A1, 15 October 1997.

Mecrow, B.C., Fully-pitched winding switched reluctance and stepping motor arrangements, *IEE Proceedings*, Pt B, Vol. 140, No. 1, pp. 61–70, 1993.

Mecrow, B.C., *Doubly salient reluctance machines*, US Patent No. 5,545,938, 13 August 1996.

Mese, E. and Torrey, D.A., Sensorless position estimation for variable-reluctance machines using artificial neuronal networks, *Conf. Rec. IEEE Industry Applications Society Annual Meeting*, 1997.

Michaelides, A.M. and Pollock, C., The effect of end core flux on the performance of the switched reluctance motor, *IEE Proc.-Electr. Power Appl.* Vol. 141, No. 6, pp. 308–316, November 1994.

Michaelides, A.M. and Pollock, C., Design and performance of a high efficiency 5-phase switched reluctance motor, *IEE 7th Int. Conf. Electrical Machines and Drives*, Durham, pp. 266–270, 11–13 September 1995a.

Michaelides, A.M. and Pollock, C., Short flux paths optimise the efficiency of a 5-phase switched reluctance drive, *Conf. Rec. IEEE Industry Applications Society Annual Meeting*, pp. 286–293, Orlando FL, October 1995b.

Michaelides, A.M. and Pollock, C., Modelling and design of switched reluctance motors with two phases simultaneously excited, *IEE Proc.-Electr. Power Appl.* Vol. 143, No. 5, pp. 361–370, September 1996.

Michaelides, A.M., Pollock, C. and Jolliffe, C.M., Analytical computation of the minimum and maximum inductances in single and two phase switched reluctance motors, *IEEE Transactions on Magnetics*, Vol. 33, No. 2, pp. 2037–2040, March 1997.

Miller, T.J.E., US Patent No. 4,500,824, *Method of commutation and converter circuit for switched reluctance motors*, 19 February 1985a (zero-volt loop).

Miller, T.J.E., Converter volt-ampere requirements of the switched reluctance motor drive, *Transactions IEEE*, Vol. IA-21, No. 5, pp. 1135–1144, 1985b.

Miller, T.J.E. and Bass, J.T., *Switched reluctance motor drive operating without a shaft position sensor*, US Patent No. 4611157, 9 September 1986.

Miller, T.J.E., Steigerwald, R.L. and Plunkett, A.B., *Regenerative unipolar converter for switched reluctance motors using one main switching device per phase*, US Patent No. 4,684,867, 1987.

Miller, T.J.E. (ed.), Switched reluctance motor drives, (collected papers) Intertec Communications Inc., Ventura, California, 1988.

Miller, T.J.E., *Brushless Permanent-magnet and Reluctance Motor Drives*, Oxford University Press, 1989 and 1993.

Miller, T.J.E. and McGilp, M., Nonlinear theory of the switched reluctance motor for rapid computer-aided design, *IEE Proceedings*, Vol. 137, Pt B, pp. 337–347, November 1990.

Miller, T.J.E., *Switched Reluctance Motors and their Control*, Oxford University Press and Magna Physics Publications, ISBN 0-19-859387-2 (UK); 9 780198593874 (USA), 1993.

Miller, T.J.E., Faults and unbalance forces in the switched reluctance motor, *Transactions IEEE Industry Applications Society*, Vol. 31, No. 2, pp. 319–328, March/April 1995.

Miller, T.J.E., Kjaer, P.C. and Schweitzer, U., Dynamic performance of the switched reluctance motor, *Proceedings of Conference on Power Conversion and Intelligent Motion*, Nurnberg, pp. 321–330, 1995.

Miller, T.J.E., Glinka, M., Cossar, C., Gallegos-Lopez, G., Ionel, D. and Olaru, M., Ultra-fast model of the switched reluctance motor, *Conf. Rec. IEEE Industry Applications Society Annual Meeting*, St Louis, USA, pp. 319–326, October 1998.

Miller, T.J.E. and McGilp, M.I., *PC-SRD* User's manual, Version 7.0, SPEED Laboratory, University of Glasgow, 1999.

Moallem, M. and Ong, C.M., Predicting steady-state performance of a switched reluctance machine, *IEEE Transactions on Industry Applications*, Vol. IA-27, No. 6, pp. 1087–1097, 1991.

Multon, B. and Glaize, C., Optimisation du dimensionnement des alimentations des machines a reluctance variable, *Revue Phys. Appl.* 22 (1987), pp. 339–357, May 1987.

Moreira, J.C., Torque minimization in switched reluctance motors via bicubic spline interpolation, *IEEE Power Electronics Specialists Conference Record*, Toledo, Spain, pp. 851–856, 1992.

Mvungi, N.H., Lahoud, M.A. and Stephenson, J.M., A new sensorless position detector for SR drives, *Proc. 4th Int. Conf. on Power Electronics and Variable Speed Drives*, pp. 249–252, 1990.

Mvungi, N.H. and Stephenson, J.M., Accurate sensorless rotor position detection in an SR motor, *Proceedings of European Power Electronics Conference*, Firenze, pp. 390–393, 1991.

Nasar, S.A., DC switched reluctance motor, *IEE Proceedings*, Vol. 116, No. 6, pp. 1048–1049, 1969.

Nasar, S.A., Boldea, I. and Unnewehr, L.E., *Permanent Magnet, Reluctance and Self-synchronous Motors*, CRC Press, 1993.

Obradovic, I.J., *Control apparatus and method for operating a switched reluctance motor*, US Patent No. 4777419, 11 October 1988.

Oldenkamp, J.L., *Reversible switched reluctance motor operating without a shaft position sensor*, US Patent No. 5440218, 8 August 1995.

Omekanda, A.M., Broche, C. and Renglet, M., Calculation of the electromagnetic parameters of a switched reluctance motor using an improved FEM-BIEM – application to different modesls for torque calculation, *IEEE Transactions on Industry Applications*, Vol. 33, No. 4, pp. 914–918, July/August 1997.

Orthmann, R., *Verfahren zum Ansteuern eines Reluktanzmotors*, EPO 0 599 334 A2, 26 November 1992.

Orthmann, R. and Schoner, H.P., Turn-off angle control of switched reluctance motors for optimum torque output, *Proceedings of European Conference on Power Electronics and Applications*, pp. 20–25, 1993.

Palaniappan, R., Dhyanchand, J.P. and Unnewehr, L.E., *Unipolar converter for variable reluctance machines*, US Patent No. 5,084,862, 28 January 1992.

Panda, S.K. and Amaratunga, G.A.J., Comparison of two techniques for closed-loop drive of VR step motors without direct rotor position sensing, *IEEE Transactions on Industrial Electronics*, Vol. 38, No. 2, pp. 95–101, April 1991a.

Panda, S.K. and Amaratunga, G.A.J., Analysis of the waveform-detection technique for indirect rotor-position sensing of switched reluctance motor drives, *IEEE Transactions on Energy Conversion*, Vol. 6, No. 3, pp. 476–483, September 1991b.

Panda, S.K. and Amaratunga, G.A.J., Waveform detection technique for indirect rotor-position sensing of switched reluctance motor drives: I. Analysis, *IEE Proceedings*, Part B, Vol. 140, No. 1, pp. 80–88, January 1993a.

Panda, S.K. and Amaratunga G.A.J., Waveform detection technique for indirect rotor-position sensing of switched reluctance motor drives: II. Experimental results, *IEE Proceedings*, Part B, Vol. 140, No. 1, pp. 89–96, January 1993b.

Panda, S.K. and Dash, P.K., Application of nonlinear control to switched reluctance motors; a feedback linearisation approach, *IEE Proc.-Elec. Power Appl.*, Vol. 143, No. 5, pp. 371–379, September 1996.

Pengov, W.A., *Staggered pole switched reluctance motor*, US Patent No. 5,852,334, 22, December 1998.

Pengov, W.A. and Weinberg, R.L., Novel shape unlocks two-phase switched reluctance motor potential, *Drives and Controls Conference*, Telford, England, 16–18 March, 1999.

Phillips, D.A., Switched reluctance drives: new aspects, *IEEE Trans. Power Electronics*, Vol. 5, No. 4, pp. 454–458, 1990.

Pollock, C. and Williams, B.W., A unipolar converter for a switched reluctance motor, *IEEE Transactions on Industry Applications*, Vol. IA-26, No. 2, pp. 222–228, March 1990a.

Pollock, C. and Williams, B.W., Power converter circuits for switched reluctance motors with minimum number of switches, *IEE Proceedings*, Pt B, Vol. 137, No. 6, pp. 373–384, 1990b.

Pollock, C. and Williams, B.W., *Reluctance motor drive system*, UK Patent GB 2,208, 456 B, 16 October 1991.

Pollock, C. and Michaelides, A.M., Switched reluctance drives: a comparative evaluation, *IEE Power Engineering Journal*, Vol. 9, No. 6, pp. 257–266, December 1995.

Pollock, C. and Wu, C.Y., Acoustic noise cancellation techniques for switched reluctance drives, *IEEE Transactions on Industry Applications*, Vol. 33, No. 2, pp. 477–484, March/April 1997.

Pollock, C., Low cost switched reluctance drives, *Transactions of the Institute of Marine Engineers*, Vol. 110, Part 4, pp. 195–205, 1998.

Pollock, C. and Barnes, M., *Switched reluctance electric machine system*, US Patent 6,037,740, 14 March 2000.

Post, R. C., *Physics, Patents and Politics – A biography of Charles Grafton Page*, Science History Publications, New York, 1976.

Press, W.H., Teukolsky, S.A., Vetterling, W.T. and Flannery, B.P. *Numerical Recipes in C*, 2nd Edition, Cambridge University Press, ISBN 0-521-43108-5, 1995.

Preston, M.A. and Lyons, J.P., A Switched Reluctance Motor Model with Mutual Coupling and Multi-Phase Excitation, *Transactions on Magnetics*, Vol. 27, No. 6, November 1991.

Pulle, D.W.J., Computation of torque in doubly salient reluctance machines using cubic splines, *IE Aust. & IREE Aust. Journal*, Vol. 8, No. 4, pp. 238–242, 1988.

Pulle, D.W.J., New data base for switched reluctance drive simulation, *IEE Proceedings, Pt B Electric Power Applications 1991*, Vol. 138, No. 6, pp. 331–337, November 1991.

Rabinovici, R., Scaling of switched reluctance motors, *IEE Proc.-Electr. Power Appl*, Vol. 142, No. 1, January 1995.

Radun, A.V., *Linearizer for a switched reluctance generator*, US Patent No 5204604, April 1993.

Radun, A.V., Generating with the switched reluctance motor, *Proceedings of IEEE Applied Power Electronics Conference*, pp. 41–46, 1994.

Radun, A.V., Design considerations for the switched reluctance motor, *IEEE Transactions on Industry Applications*, Vol. 31, No. 5, September/October 1995.

Radun, A.V., Ferreira, C.A. and Richter, E., Two-channel switched reluctance starter/generator results, *IEEE Transactions on Industry Applications*, Vol. 34, No. 5, pp. 1026–1034, September/October 1998.

Rao, S.C., *AC synchronous motor having an axially laminated rotor*, US Patent No. 4110646, August 1978.

Ray, W.F. and Davis, R.M., Inverter drive for doubly salient reluctance motor; its fundamental behaviour, linear analysis, and cost implications. *IEE Proc.-Electr. Power Appl.*, Vol. 2, pp. 185–193, 1979.

Ray, W.F. and Davis, R.M., *Reluctance electric motor drive systems*, US Patent No. 4,253,053, 24 February 1981. Also UK Patent No. 1,591,346, 17 June 1981 (current waveform).

Ray, W.F., *et al.*, High performance switched reluctance brushless drives, *Transactions IEEE*, Vol. IA-22, No. 4, pp. 722–730, July/August 1986.

Ray, W.F. and Al-Bahadly, I.H., Sensorless methods for determining the rotor position of switched reluctance motors, *Proceedings of European Power Electronics Conference*, pp. 7–13, 1993.

Ray, W.F. and Al-Bahadly, I.H., A sensorless method for determining rotor position for switched reluctance motors, *IEE Conf. on Power Electronics and Variable Speed Drives*, pp. 13–17, October 1994.

Ray, W.F., *Sensorless rotor position measurement in electric machines*, US Patent No. 5467025, 14 November 1995.

Reay, D.S., Green, T.C. and Williams, B.W., Neural networks used for torque ripple minimisation from a switched reluctance motor, *Proceedings of European Conference on Power Electronics*, Brighton, pp. 1–7, 1993a.

Reay, D.S., Green, T.C. and Williams, B.W., Application of associative memory neural networks to the control of a switched reluctance motor, *IEEE-IECON Conf. Record*, Maui, pp. 200–206, 1993b.

Reece, A.B.J. and Preston, T.W., *Finite Element Methods in Electrical Power Engineering*, Oxford University Press, 2000.

Regas, K.A. and Kendig, S.D., Step-motors that perform like servos, *Machine Design*, pp. 116–120, 10, December 1987.

Reichard, J.G. and Weber, D.B., *Switched reluctance electric motor with regeneration current commutation*, US Patent application C.7566–2306, 23 May 1989.

Reinert, J., Smith, E.D. and Enslin, J.H.R., Drive of a high torque, low speed switched reluctance machine, *Conf. Rec. IEEE Industry Applications Society Annual Meeting.*, pp. 259–264, 1991.

Reinert, J., Inderka, R., Menne, M. and De Doncker, R., Optimizing performance in switched reluctance drives, *13th Annual Applied Power Electronics Conference and Exposition APEC-98*, Anaheim, CA, pp. 765–770, 15–19, February 1998.

Richter, E. and Ferreira, C.A., Performance evaluation of a 250 kW switched reluctance starter generator, *Conf. Rec. IEEE Industry Applications Society Annual Meeting*, pp. 434–440, 1995.

Richter, E., High temperature, lightweight switched reluctance motors for future aircraft engine applications, *Proc. American Control Conf.*, Atlanta, pp. 1846–1851, June 1988.

Richter, E., Anstead, D.H., Bartos, J.W. and Watson, T.U., Preliminary design of an internal starter/generator for application in the F110-129 engine, SAE Aerospace Atlantic Conference, Dayton, SAE paper No. 951406, pp. 1–10, 23–25 May 1995.

Ritchie, W., Experimental researches in electro-magnetism and magneto-electricity, *Phil. Trans*, [2] pp. 313–321, 1833.

Rossi, C. and Tonielli, A., Feedback linearising and sliding mode control of a variable reluctance motor, *International Journal of Control*, Vol. 60, No. 4, pp. 543–568, 1994.

Russa, K., Husain, I. and Elbukuk, M.E., Torque-ripple minimization in switched reluctance machines over a wide speed range, *IEEE Transactions on Industry Applications*, Vol. 34, No. 5, pp. 1105–1112, September/October 1998.

Sahoo, N.C., Xu, J.X. and Panda, S.K., Determination current waveforms for torque ripple minimization in switched reluctance motors using iterative learning: an investigation, *IEE Proc.-Electr. Power Appl*, Vol. 146, No. 4, pp. 369–377, July 1999.

Sakano, T, *Circuit and method for driving switched reluctance motor*, WO 94/08391 PCT patent specification, 14 April 1994.

Sawata, T., Kjaer, P.C., Cossar, C., Miller, T.J.E. and Hayashi, Y., Fault-tolerant operation of single-phase switched reluctance generators, *IEEE Transactions on Industry Applications*, Vol. 35, No. 4, pp. 774–781, July/August 1999a.

Sawata, T., Kjaer, P.C., Cossar, C. and Miller, T.J.E., A study on operation under faults with the single-phase SR generator, *IEEE Transactions on Industry Applications*, Vol. 35, No. 4, pp. 782–789, July/August 1999b.

Sawata, T., Kjaer, P.C., Cossar, C. and Miller, T.J.E., A control strategy for the switched reluctance generator, *ICEM '98*, Istanbul, Vol. 3, pp. 2131–2136, 2–4 September 1998.

Sawata, T., Yamaguchi, Y. and Miller, T.J.E., Analysis of an optimum number of the multi-tooth per pole SRM, *IPEC 2000*, Tokyo, Japan, Vol. 1, pp. 637–642, 3–7 April 2000.

Schramm, D.S., Williams, B.W. and Green, T.C., Optimum commutation-current profile on torque linearization of switched reluctance motors, *Proceedings of International Conference of Electrical Machines*, Manchester, pp. 484–488, 1992.

Sood, P.K., *Power converter for a switched reluctance motor*, US Patent No. 5,115,181, 19 May 1992.

Sood, P.K., Skinner, J.L. and Petty, D.M., *Method and apparatus of operating a dynamoelectric machine using DC bus current profile*, US Patent No. 5420492, 30 May 1995.

Soong, W.L. and Miller, T.J.E., Theoretical limitations to the field-weakening performance of five classes of brushless synchronous AC motor drive, *Proceedings of Electrical Machines and Drives Conference*, pp. 127–132, Oxford, 1993.

Stephens, C.M., Fault detection and management system for fault tolerant switched reluctance motor drives, *IEEE Transactions on Industry Applications*, Vol. 27, No. 6, pp. 1098–1102, November/December 1991.

Stephens, C.M. and Radun, A.V., *Current chopping strategy for generating action in switched reluctance machines*, US Patent No. 5166591, 24 November 1992.

Stephenson, J.M. and Corda, J., Computation of torque and current in doubly-salient reluctance motors from nonlinear magnetization data, *IEE Proceedings*, Vol. 126, No. 5, pp. 393–396, 1979.

Stephenson, J.M. and El-Khazendar, M.A., Saturation doubly-salient reluctance motors, *IEE Proceedings*, Vol. 136, Pt B, No. 1, pp. 50–58, January 1989.

Stephenson, J.M., *Improvements in electric reluctance machines*, European Patent Specification, EP 0 414 507 A1, 21 August 1990 (projection increases contact area).

Stephenson, J.M., *Improvements in electrical machines*, European Patent Specification, EP 0 343 845 B1, 1993.

Stephenson, J.M., *Switched reluctance motors*, European Patent Specification, EP 0 601 818 B1, 1996.

Stephenson, J.M., Computer-optimized smooth-torque current waveforms for switched reluctance motors, *IEE Proc.-Electr. Power Appl*, Vol. 144, No. 5, pp. 310–316, 1997.

Stephenson, J.M. and Jenkinson, G.C., Single-phase switched reluctance motor design, *IEE Proc.-Electr. Power Appl.*, Vol. 147, No. 2, pp. 131–139, March 2000.

Stiebler, M. and Ge, J., A low-voltage switched reluctance motor with experimentally optimized control, *Proceedings of International Conference of Electrical Machines*, Manchester, pp. 532–536, 1992.

Sturgeon, W., Improved electro-magnetic apparatus, *Trans. Soc. Arts, Manufactures & Commerce*, Vol. XLIII, pp. 37–52, plates 3 & 4, 1824.

Suriano, J. and Ong, C.-M., Variable reluctance motor structures for low speed operation, *Conf. Rec. IEEE Industry Applications Society Annual Meeting*, Toronto, pp. 114–121, 2–8 October 1993.

Takahashi, T., Chiba, A., Ikeda, K. and Fukao, T., A comparison of output power control methods of switched reluctance motors, *IEEE-PCC Conf. Record*, pp. 390–395, 1993.

Takayama, K., Takasaki, Y., Ueda, R., Sonoda, T. and Iwakane, T., A new type switched reluctance motor, *Conf. Rec. IEEE Industry Applications Society Annual Meeting*, Pittsburgh, pp. 71–78, 2–7 October 1988.

Takayama, K., *et al.*, Thrust force distribution on the surface of stator and rotor poles of switched reluctance motor, *IEEE Transactions on Magnetics*, Vol. MAG-25, No. 5, pp. 3997–3999, September 1989.

Tandon, P., Rajarathnam, A.V. and Ehsani, M., Self-tuning control of a switched-reluctance motor drive with shaft position sensor, *IEEE Transactions on Industry Applications*, Vol. 33, No. 4, pp. 1002–1010, July/August 1997.

Török, V., *Improvements in electrical inductor machines*, UK Patent No. 1,500,798, 19 April 1974.

Török, V., *Electrical reluctance machine*, US Patent No. 4,349,605, 1982.

Török, V. and Loreth, K., The world's simplest motor for variable speed control? The Cyrano Motor, A PM-biased SR-motor of high torque density, European Power Electronics Conference, Brighton, UK, 13 September 1993.

Torrey, D.A. and Lang, J.H., Modelling nonlinear variable reluctance motor drive, *Proceedings IEE*, Vol. 137, Pt B, No. 5, pp. 314–326, 1990.

Torrey, D.A. and Hassanin, M., The design of low-speed variable-reluctance generators, *Conf. Rec. IEEE Industry Applications Society Annual Meeting*, pp. 427–433, 1995.

Torrey, D.A., An experimentally verified variable-reluctance machine model implemented in the SABER circuit simulator, *Electric Machines and Power Systems*, No. 24, 1996.

Unnewehr, L.E. and Koch, W.H., An axial air-gap reluctance motor for variable-speed applications, *IEEE Transactions*, Vol. PAS-93, pp. 367–76, 1974.

Van der Broeck, H., DE 3819097 German Patent, 4 June 1988.

Van Sistine, T.G., *Apparatus for starting a switched reluctance motor*, US Patent No. 5497064, 5 March 1996a.

Van Sistine, T.G., *Switched reluctance motor providing rotor position detection at low speeds without a separate rotor shaft position sensor*, US Patent No. 5525887, 11 June 1996b.

Van Sistine, T.G., Bartos, R.P., Melhorn, W.L. and Houle, T.H., *Switched reluctance motor providing rotor position detection at high speeds without a separate rotor shaft position sensor*, US Patent No. 5537019, 16 July 1996.

Vitunic, M., *Switched reluctance motor controller with sensorless rotor position detection*, US Patent No. 5859518, 12 January 1999.

Vukosavic, S., Peric, L., Levi, E. and Vuckovic, V. Sensorless operation of the SR motor with constant dwell, *Proc. IEEE Power Electronics Specialists' Conference*, pp. 451–554, 1990.

Wallace, R.S. and Taylor, D.G., Low torque ripple switched reluctance motors for direct-drive robotics, *IEEE Transactions on Robotics and Automation*, Vol. 7, No. 6, pp. 733–742, 1991.

Wallace, R.S. and Taylor, D.G., A balanced commutator for switched reluctance motors to reduce torque ripple, *IEEE Transactions on Power Electronics*, Vol. 7, No. 4, pp. 617–626, 1992.

Warne, D.F. and Calnan, P.G., Generation of electricity from the wind, *IEE Proceedings*, Vol. 124, No. 11R, pp. 963–985, November 1977.

Watkins, S.J., *Sensorless rotor position monitoring in reluctance machines*, US Patent No. 5793179, 11 August 1998.

Welburn, R., *Ultra-high torque motor system for direct drive robotics*, Motor-Con Proceedings, Atlantic City, pp. 17–24, April 1984.

Weller, A., DE 40 36 565 C1, 16 November 1990.

Wu, C.Y. and Pollock, C., Analysis and reduction of vibration and acoustic noise in the switched reluctance drive, *IEEE Transactions on Industry Applications*, pp. 91–98, January/February 1995.

Wu, C.Y. and Pollock, C., *Electric motor drive*, US Patent 5,767,638, 16 June 1998.

Xu, L. and Bu, J., Position transducerless control of a switched reluctance motor using minimum magnetizing input, *Conf. Rec. IEEE Industry Applications Society Annual Meeting*, 1997.

Index